Quality Concrete from Crap

*Production techniques
used to create quality concrete
from less-than-ideal
materials*

By: Herb Nordmeyer

*Foreword by: Renaud Bury,
Dean of the College of Engineering,
American University of the Caribbean,
Les Cayes, Haiti*

*Published by:
Nordmeyer, LLC
Castroville, Texas, USA   78009
2017*

Copyright Page

| **ISBN:** | English Print Version | 978-0-9960100-8-5 |
| | Haitian Creole Print Version | 978-0-9960100-9-2 |

Library of Congress Control Number: 2017917995

Author portrait by:	Yolanda Chapin, Photographic Memories http://yolandachapin.com
Cover design by:	Deborah M. Salinas Shepherd of the Hills Lutheran School San Antonio, TX
Illustrations by:	Author & Mission:Haiti mission teams

Library of Congress Subject Headings:
Construction
Concrete
Mortar
Stucco

Trademarks:

Easy-Spred is a registered trademark of Peninsular Products, Inc.
Gunite is a registered trademark of Allentown Equipment Co.
Kel-Crete is a registered trademark of Kel-Crete Industries, Inc.
Styrofoam is a registered trademark of Dow Chemical Co.

Published by:
Nordmeyer, LLC
213 County Road 575
Castroville, Texas 78009-2120
www.NordyBooks.com

Acknowledgements

Rev. Dr. Paul M. Touloute, Provost of the American University of the Caribbean, and Renaud Bury, Dean of the College of Engineering of the American University of the Caribbean, have both encouraged me to write this book and have provided assistance. They have also arranged for my teaching at the American University of the Caribbean.

James Hicks, Paul Taylor, Craig Cannon, and Rick Garagliano provided numerous editorial suggestions. In fact, between the 4 of them they provided over 2,500 suggestions.

David South and the Monolithic staff have stood behind me, given me suggestions and encouragement, and inspired me to do more than I am capable of doing.

Lophane Laurent and his team ensure that we have whatever we need in Haiti and that we remain safe. They also point out when we speak in a manner which people do not understand.

Debbie Salinas is an excellent artist who designs my book covers, draws pictures for some of my books, and constantly encourages me to continue working.

Quality Concrete from Crap

Helen Roenfeldt is the Executive Director of Mission: Haiti. She arranges trips, discovers things which need to be done, and picks up the pieces when I fail to pack all the supplies which we need on a project.

Judy Nordmeyer is my wife and proofreader. She keeps the home fires burning when I am gallivanting around the world. Even though her health will not allow her to travel with me, she is constantly ensuring that I can travel at a moment's notice.

The many who laughed when they heard that I was working on this book, because they knew it was impossible, and the many more who laughed each time I mumbled that I was going to drop the project because it was impossible.

Table of Contents

Quality Concrete from Crap

Table of Contents

Quality Concrete from Crap

Table of Contents

Quality Concrete from Crap

Table of Contents

Foreword

The gems that Dr. Herb Nordmeyer, he prefers to be called Herb by young and old alike, shared with the community of the American University of the Caribbean (AUC), in Les Cayes, Haiti, during his January 2017 visit are both priceless and memorable. Many of us continue building upon the principles and insights he shared with us.

Yet Herb had a strong desire to continue helping not only the AUC community but the Haitian community at large—especially those who did not have the opportunity to attend his presentations. Thus, he left the AUC with the determination to combine his presentations into a book that could be read by countless people in the construction field and so contribute to the use of better construction practices in Haiti.

Herb's book *Quality Concrete from Crap*, which scrutinizes construction failure in the South and Grand'Anse Departments after Hurricane Matthew, will prove instrumental in our preparedness for the next natural disaster since Haiti is exposed to unpredictable natural disasters, such as earthquakes and hurricanes. Knowing that the quality of materials used plays a key role in the life span of a construction and considering the economic limitations of many in the population, the author presents some great alternatives to such expensive materials as asphalt and concrete.

The author encourages the improvement of traditional methods and use of local materials to build habitats that protect the people's investments and lives from potential disasters. He sees endless value in the Caribbean construction style that can be sold to the entire world because it is so natural but as yet un-discovered by most of the world.

Quality Concrete from Crap is written in a simple and clear language which makes it understandable and useful even to the least-educated construction crew member in the remotest countryside. It is a priceless treasure that Herb has bestowed upon the Haitian people.

Even though he prefers Herb, I will continue, and encourage others, to call him Dr. Nordmeyer

Renaud BURY,
Dean of the College of Engineering,
American University of the Caribbean,
Les Cayes, Haiti
Phone number: (+509) 4868-1715/ 4331-7541
E-mail: renaudburyboom@yahoo.fr

Preface

Formatting

Inserting photos and figures in a book leads to problems. Some publishers move text around so there are no blank spots on pages. This separates the photos from the text concerning them. Some reduce the size of the photos, so everything fits, even though the photos are so small that they cannot be examined. We have chosen to leave large blank spaces on some pages to eliminate the above problems.

Purpose—Why was this book written?

Just as any structure requires a good foundation, so does a discussion about quality concrete. As such, it is important to define the conditions this book was written for. Many books contain this type of explanation, but hide it at the end. Because the purpose of this book is to assist people of various education levels who speak different languages, we have chosen to establish this foundation up front.

If this were a scientific paper or a book on concrete for people knowledgeable about quality concrete, it would include much more detail and would reference numerous published procedures and standards. This book is being written for people who are building homes, either as contractors or for themselves, and do not have access to modern scientific equipment and to materials which are certified as meeting ASTM standards. It is designed to help them build disaster-resistant homes.

While the techniques discussed in this book can be used for many different uses of concrete, the over-riding use discussed in this book is disaster-resistant housing. There are too many examples of houses built with good intentions that have failed with the next hurricane or within five years, whichever comes first.

Concrete—What is it?

For this book, concrete is defined as a substance which contains cement paste and aggregate. That means that it includes, but is not limited to:

> Ordinary Concrete
> Stucco
> Mortar
> Ferro-cement
> Surface Bonding Cement

While there are differences between the various products, they all have many common characteristics.

The cement paste may be:

> Portland cement and water, or
> Pozzolanic materials, hydrated lime, and water, or
> Other materials which will harden under water and will bond aggregate together, or
> A combination of the above.

In this book, the discussion is limited to Portland cement and water paste, but one should recognize that Haiti and many other countries have the raw materials to produce a pozzolanic/lime cement. Haiti does not have the fuel to do so.

5

Scope

This book was originally designed for developing quality concrete in Haiti and was expanded to cover developing quality concrete worldwide. As such, it will discuss materials that may not be available in each part of the world. For example, at the time of this book's writing, Haiti does not have the raw materials to produce a pozzolanic/lime cement. However, pozzolanic/lime options are discussed because it could be an option in other areas now or in Haiti in the future.

Definitions

A complete glossary is in the back of the book. These terms are defined in the preface to allow the reading of this book with as few problems as possible.

Aggregate: Aggregates should be inert granular materials such as sand, gravel, or crushed stone and are usually sized through screens to specific sizes. Optimally, the amount of each component should be such that the resulting aggregate will have a minimal pore space and thus will be as dense as possible.

Cement Paste: Any material(s) and water that when mixed will bond aggregate together. Two common mixtures that form cement paste are Portland cement & water and pozzolanic materials, hydrated lime, & water.

Chlorides: (commonly referred to as salts) Chlorides occur when the chlorine element gains an extra electron (written as Cl-). Chlorides increase corrosion of metal rebar, thus creating a fault within the concrete. Chlorides are commonly introduced into concrete through the water used to make concrete paste; however,

Quality Concrete from Crap

6

the soil in the area may contain high levels of
chlorides as well.

Compression: One of the two key types of force
involved in building (the other is Tension).
Compression is a force that squeezes the material. Concrete has a high compression strength.

Fines: Sand-sized material—from 0.25 to 3.2 mm.

Hydrated Lime: Calcium hydrate. It is formed by
calcining limestone and then adding measured
amounts of water to hydrate the resulting calcium oxide. Normally, it is classified as Type S
(dolomitic and highly hydrated), Type N (high
calcium and not as highly hydrated), and industrial grade (not as highly hydrated and the
particle size is not as controlled).

Portland Cement: Any of several grades of cement
which meet ASTM Standard C150. This is the
most common cement on the market.

Pozzolan: A non-crystalline and finely divided
amorphous (non-crystalline) silicate or aluminate which in the presence of calcium ions can
form hydrated calcium silicates and aluminates
(hydrated cement chemicals).

Surface Bonding Cement: A stucco-like material
with a high bond strength and reinforcing, so
concrete block and other masonry units can be
dry-stacked and then bonded by plastering the
surface.

Tension: One of the two key types of force involved
in building (the other is Compression). Tension

Preface

is the force that pulls materials apart. Concrete has a low tensile strength, which is why reinforcing materials are added (i.e. rebar).

Ultrafines: Extremely small (less than 0.25 mm down to smaller than a cement particle) materials that are intentionally or unintentionally part of the concrete mix. Some ultrafines (such as pozzolanic material) strengthen concrete; other ultrafines (such as non-pozzolanic clay) weaken concrete.

Chapter 1

Prologue

How Long Do You Want Your Concrete to Last?

At the beginning of any project involving concrete, one must ask the most important question in the process, "How long do I want the concrete to last?" This is the question, because the answer will inform all the decisions that will come after it. There are many possible answers to the question, and they are all based upon a lot of factors.

> Until I get paid for the job?
> Until the next hurricane?
> Until the next earthquake?
> Until my grandchildren are old and gray?

Only you can determine that answer and make it happen. And the answer will have far greater effects on one's business than just that one project.

If you only care about getting paid, you can cut corners, but it will have long-term effects, both personally and professionally. How will you feel if the home you build for a client fails within the next five years? Or if it collapses on a child and kills her?

If you sacrifice quality for short-term business, as the years pass you will find it more difficult to get clients, especially if your competitor has adopted a quality mind-set.

If your structure fails in the next hurricane or earthquake, but your competitor's structure survives, how will that affect future business? If you sacrifice quality, you are going to sacrifice future business growth.

If you build quality structures, you may struggle at first to get clients because they may not be able to envision a home that is likely to survive the next disaster.

If you opt for quality, you may need to explain to your clients why it is important that they pay a little more than what your competitor is charging. Many clients may not be able to envision structures that last, having experienced only bad concrete. You need to be ready to explain, in words that your client understands, why it is important.

While this is about the shortest chapter in this book, it is probably the most important. If you do not opt for quality, there is little reason for reading the rest of the book.

Chapter 2

Concrete Physics

Drying Shrinkage

Concrete, stucco, and mortar are at their largest volume either as they are placed or shortly thereafter, when the cement hydration starts to heat the mix. As the mix cures, water is lost, and the volume shrinks. On a sidewalk without control joints, there is usually a crack every seven steps. Either the contractor determines where the crack will be by inserting a control joint, or the concrete will make that determination.

If steps are not taken to slow down the drying of concrete, the surface will dry faster, shrink, and cause surface cracks. Since the top of the slab is shrinking and the bottom of the slab is not, the slab can curl. This can result in a portion of the slab not supported by base material. These are places where structural cracks can start to form. They are also places where chlorides and sulfates can get into the core of the concrete and potentially enhance the degradation of the steel rebar or the concrete.

Thermal Movement

When concrete gets hot, it expands. When it cools off, it shrinks. If sun shines on a concrete slab or on the western wall of a stucco or concrete building, it can get quite hot and cause cracks. Remember the sidewalks and seven steps.

Moisture Movement

It may take a year or more for excess moisture to work its way to the surface of a slab. The more air voids, the faster the moisture will move. To prevent moisture from moving from the soil up through the concrete, a barrier is needed or the source of the moisture must be eliminated.

The moisture itself may not be a problem; but if the soil contains high levels of chlorides or sulfates, they may cause the concrete to deteriorate. An additional problem is if people sleep on the floor or if their bedding is on the floor, moisture wicking through the slab can add moisture to their bedding, and this can result in mold growth. To keep ground moisture from wicking up, a sheet of polyethylene plastic is often placed under the slabs and footings.

Concrete Strength

Concrete has excellent compressive strength and poor tensile strength. Steel has excellent tensile strength but is expensive. Stresses always occur on concrete.

When a roof sags, the cracks start on the underside, where the concrete is under tension. It also starts on the top where there is support below.

Simple Roof Slab

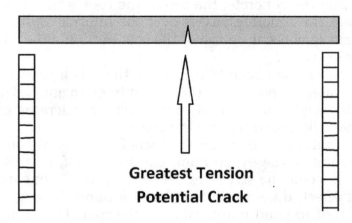

Greatest Tension
Potential Crack

Figure 1. Simple roof. Concrete is heavy; and where it is not supported, tension develops on the underside. This can lead to cracks. Reinforcement is installed to handle the tension.

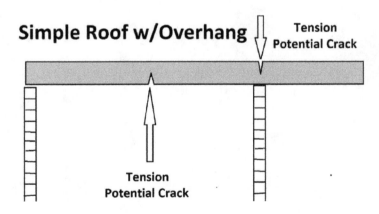

Simple Roof w/Overhang

Tension
Potential Crack

Tension
Potential Crack

Figure 2. Simple roof with overhang. When there is an overhang from a wall, tension develops above the wall, as well as in unsupported areas.

Quality Concrete from Crap

Concrete roofs are seldom as simple as the two drawings. If there are internal walls which are supporting the concrete, the top of the roof is in tension where the concrete passes over the internal walls, so cracks may occur on top of the roof outlining the supporting walls.

Since the concrete roof sags, there is a tendency for water to pool over those points of sagging. This adds weight to the roof, and it also accelerates the penetration of water into the concrete.

That water does have a beneficial use. When it evaporates, the evaporation cools the roof, and as a result, cools the structure below the roof. To enhance this effect, dikes are often placed around the edge of the roof to pond more water on the roof. This results in water standing over the tension cracks which occur above the load-bearing walls. This accelerates the deterioration of steel rebar that is used in the roof slab.

Chapter 3
Why Does Concrete Fail?

Photos of Concrete Failures

Following are several photos of concrete failures which have occurred within a few years of the structure being built. Each builder should ask himself, "Do I want my children and grandchildren to see that this is the way I build?"

Try to identify the cause of each failure. Remember, there is often more than one cause.

Photo 1

Photo 2

Why Does Concrete Fail?

Photo 3

Quality Concrete from Crap

Photo 4

Why Does Concrete Fail?

Photo 5

Quality Concrete from Crap

Photo 6

Why Does Concrete Fail?

Photo 7

Quality Concrete from Crap

Photo 8

Why Does Concrete Fail?

Explanation of Concrete Failures

Now, here are some answers.

Photo 1. Crumbling ceiling. Steel rebar corroded and expanded. This caused the lower layer of the ceiling/roof to break off. Additionally, the block used as space fillers in the ceiling/roof were weak, and they broke. Additional problems included water leaking through the roof, low strength concrete, and unconsolidated concrete being used.

Photo 2. Deteriorated column. This column was poured about 5 years before the photo was taken. The aggregate used had a high clay, content and the amount of Portland cement used was substantially below the amount needed to hold the concrete together.

Photo 3. Deteriorated fence. An attempt was made to build a fence using confined masonry. The column is not attached to the fence panel. The concrete in the column is not consolidated. The beam along the top of the wall was undersized, and the concrete in the beam is eroding because of low cement content and excessive fines in the aggregate.

Photo 4. Disaster-resistant home. In Léogâne four homes were built which were certified as being disaster-resistant. All four were destroyed by Hurricane Matthew in 2016. This one was destroyed worse than the other three. Among other things, the stucco used on the lath was high in ultrafines and could be crumbled with

Quality Concrete from Crap

light finger pressure. The lath was applied to wood framing with short nails which were driven in partially and then bent over. The concrete floor was eroded and in some cases completely gone because of the low cement content and high ultrafines content. An adjacent shack survived without damage. Disaster-resistant certification means nothing unless it is built according to the plans and specifications.

Photo 5. *Roofless home. In Kenscoff, this home lost its metal roof because the bond beam to which the roof was attached was made with concrete in which the sand had a high clay content and the amount of Portland cement was less than required.*

Photo 6. *Failed column. A downdraft caused by Hurricane Matthew depressed the roof of a Catholic Church. This resulted in lateral pressure which cracked one wall and cracked one of the columns in that wall. While the concrete work appeared to be well done, after the wall and column cracked, an examination of the column revealed that it was not reinforced.*

Photo 7. *Everything went wrong Sometimes it is easier to comment on what was done correctly in a photo rather than trying to comment on all the errors. First, the drainage around the home allowed water to get close to the foundation and wash the footings away. The slab was poured without reinforcing. The column in the front of the photo was buried in the ground, but that ground washed away. The roof slab has a dip in it, indicating that it was not*

Why Does Concrete Fail?

poured flat or that it has since settled. The walls have washed away.

Photo 8. Deteriorating wall. The mortar holding this foundation has eroded. This erosion was caused by water flowing down the wall, and by the concrete having been made with an excessive amount of clay and not enough Portland Cement.

Common Causes for Concrete Failure

Here is a little more information about the causes of failure that have been mentioned.

Poor Mix

Poor Mix is a common answer, but it means nothing because it covers everything. With this book and the course based on this book, poor mix is not an acceptable answer. Why is the mix poor? Was there too much water? Was there a shortage of Portland cement? Was there too much clay in the mix? Was the concrete not consolidated? Was the aggregate soft? Were chlorides incorporated into the mix? Was the galvanized, painted, or coated rebar dragged to the job site on a motorbike, resulting in the coating being scraped off? These are just a few of the conditions that can be lumped into an answer of "poor mix," but contains no information which can be useful in attempting to prevent the problem in the future.

Poor Water/Cement Ratio

One bag of Portland cement needs 15 liters of water to fully hydrate. That is a water/cement ratio of 0.35. But there is a little problem with such a mix. It is unworkable. It can be placed in a mold and hammered into place or vibrated into place, but to handle it with normal concrete-placing tools, it is unworkable.

Adding water to make the mix workable reduces the strength of the concrete. Abram's Law, in Chapter 4, addresses the impact of water on concrete strength.

The easiest way to have enough water for handling, but not so much water that the strength of the concrete is impaired, is to use a slump cone. The use of the slump cone is addressed in Chapter 5.

Poor Air/Cement Ratio

Ferret's Law is like Abram's Law, but it factors in the absolute volume of entrained air, as well as the absolute volume of water. In fact, Ferret's Law was derived first, and Abram's Law is a simplification. Entrained air is tiny bubbles of air that are critical to the performance of the concrete. It is especially critical in areas where freeze/thaw cycles occur, but it also increases the workability of the concrete.

Non-Reactive Fines

Nordmeyer's Law is like Ferret's Law, but it factors in the absolute volume of non-reactive fines. Those fines can be clay, rock dust, soil, or anything else which is fine in nature and does not participate in the cement-curing reaction.

Poor Mixing

It takes a lot of work to mix concrete by hand. If the particles are not evenly distributed, the qualities we look for in concrete will be negatively impacted. Mechanical mixing is much easier, but sometimes there is such a rush on the job that the concrete is not adequately mixed. Another problem with mechanical mixing: if the concrete is not needed immediately, it may remain in the mixer with the mixer turning. This leads to entraining excess air.

Poor Foundations

Foundations support the building. If they are not adequate for handling the loads of the structure which will be placed on the foundation, extra stress is placed on the building. It is common that buildings are not designed to compensate for weak foundations; consequently, the building fails,

There are two major categories which must be considered when a foundation is designed and installed.

The Soil

Structures are built on earth. Some of that earth is rock, some sand, some clay. Each has qualities which must be considered when designing a foundation. Some of that earth changes size as the moisture content changes. Some loses its ability to support structures built on it as the moisture content increases. If the foundation which has been built was not built to withstand the changing conditions of the soil, the foundation will fail, and the building itself will fail.

The Concrete

If the concrete in the foundation is not designed to handle the stress that comes from the weight of the building, the wind load on the building, the snow and ice load on the building, the concrete in the foundation will fail, and the building will fail.

Poor Finishing

If the mortar between block is not tooled, water can more easily enter the wall and degrade the wall. If plaster is finished with a wet sponge so it can be leveled easier, the surface becomes porous and allows water to enter. It also produces a softer stucco, and it is more likely to erode. If a slab is wetted so a slicker finish can be developed, the surface will be softer than the concrete further down in the slab. This will cause the surface to turn to powder (dust) as it is walked on, driven on, or in other ways put into use. Besides the floor wearing out, it allows water to enter if the slab is exposed to water. Also, the more a finish is worked, the more cement paste is brought to the surface and the more hairline cracks will develop in the finish.

Poor Reinforcement

Non-reinforced concrete is strong in compression but weak in tension and flexion. Reinforcement is added to make the concrete strong in tension and flexion. Since portions of all concrete slabs, walls, and roofs are under tension, insufficient reinforcement, poor placement of the reinforcement, damaged reinforcement, or use of wrong reinforcement materials will lead to structural failure.

Deterioration of Rebar

This explanation will probably make the eyes of most readers glaze over, but for those working in an area with moderate or high chlorides, it is mandatory that the problem be addressed.

The Basics

Non-reinforced concrete is high in compressive strength, but low in tensile strength. Reinforcing is added to concrete to increase the tensile strength. The most common method used is to add steel rebar. In most situations, this is adequate. The two most common problems with the use of steel rebar are when chlorides are present, as when salt (sodium chloride) is added to a roadway for deicing, and when water can get to the rebar.

The Chemistry

When chlorides are in contact with steel, and there is enough moisture present for a chemical reaction to take place (enough moisture to dissolve some of the sodium chloride), several chemical reactions take place. First, when the salt dissolves in the water, the ability of the water to carry an electrical current is increased. This increases the ability of the oxygen that is dissolved in the water to react with iron which is in the steel rebar. In the process, ferrous oxide, ferric oxide, ferrous chloride, and ferric chloride can be formed. The oxidation state of the iron (ferric or ferrous) and whether a chloride or an oxide is formed are not germane to this issue, since each of those chemical compounds increases the volume of the iron in the steel which was the precursor for those compounds.

The Physics

When the rebar expands as these compounds are formed, stress is developed until that stress is greater than the tensile strength of the concrete. This causes the concrete to crack. The cracking of the concrete allows more water and oxygen to come in contact with the rebar and the chlorides which are present. This accelerates the process.

The stresses the concrete has been under usually weaken the concrete in areas around any of the cracks which occur.

As the chemical reactions continue, the tensile strength of the rebar decreases. This causes the tensile strength of the concrete to decrease.

When a concrete beam or slab is suspended between multiple supports, it flexes very slightly. At a point between any two supports, the bottom side of the beam is under tension and the upper side is under compression. Above any of the supports, the top of the beam is under tension and the bottom of the beam is under compression.

As a result, a flat concrete roof will often crack above any supports and will sag between those supports. The cracking above the supports increases the

moisture entering the roof and accelerates the deterioration.

Accelerants for Corrosion

In the Gonaives, Haiti, region, there are several
other factors which contribute to the problem:

The water used for mixing much of the concrete is
often brackish. This provides the chlorides to increase
the oxidation of the steel.

Much of the aggregate used contains clay. Tests
have shown up to 29% clay in concrete sand on our
jobs. The presence of clay requires more water to be
added to make the concrete, stucco, or mortar workable. The increased amount of water needed increases
the amount of chlorides which are available to catalyze the oxidation reaction.

The concrete is often porous due to aggregates
used which are not well graded to increase the density
of the concrete mass.

The concrete is often not well-consolidated, leaving air pockets in the concrete as reservoirs of oxygen
and water.

The clay in the concrete acts as a reservoir for the
chlorides, so the chlorides can initiate a chemical reaction whenever the moisture content of the concrete
is conducive to the oxidation reaction.

Conclusion

Chlorides do not directly harm the Portland cement paste or the hardened Portland cement paste.
Chlorides catalyze the oxidation of the chemical element iron, which is a major component of steel rebar
and metal lath. The oxidation process weakens and/or
destroys the tensile strength of reinforced concrete
and stucco. Expansion of the metal components can,

and often does, crack the concrete or stucco and thus accelerate the deterioration.

Avoid chlorides

Poor Attention to Detail

Pouring concrete is a complex operation, and all parts of it need to be planned carefully. Failure to take care of each detail can lead to failure of the entire job.

> *Author's Note: Nearly 40 years ago I poured concrete on a job, my father was there watching, and I did not do a final check on the forms to ensure that they were adequately braced. A little concrete seeped from close to the bottom of the forms as the concrete was added and rodded. Then the rate of seepage increased. Then a piece of plywood near the seepage came loose, and concrete started flowing out of the forms. Not only was the concrete lost, but we had to clear out the concrete before it set, remove the forms that remained, and start all over. As if I would ever forget that day and the lessons I learned, for years my father ensured that I did not forget that one day I was not prepared for the concrete pour.*

Looking Forward

In the next chapters, we will go into greater depth concerning each of these reasons which lead to concrete failure. Remember, taking care of one problem does not lead to good concrete; we need to take care of <u>all</u> the problems. This slows a job down, but consider which of these is the better deal for the customer:

Why Does Concrete Fail?

- *Building a 28-square-meter house for $6,000 US, and it lasts until the next hurricane or for 10 years, whichever comes first.*

or

- *Building a 28-square-meter house for $8,000 US, and it lasts through hurricanes and earthquakes and is there when the builder's grandchildren are old and gray.*

Following the discussion of producing good concrete, we will address some specific uses of concrete, including the building of disaster-resistant housing. We have published a book on building domes, and the Haitian Government has published a book on building disaster-resistant homes with confined masonry. Now someone needs to step forward and publish books on the other ways which concrete can be used to build disaster-resistant homes.

Chapter 4

Mix Design Considerations

Impact of Particle Size Distribution

Formula for the volume of a sphere = $4/3 \cdot 3.14 \cdot r^3$

Formula for the volume of a cube = s^3

Relatively-round gravel with all particles the same size has a void ratio of 47.7%. It does not matter if all the gravel is the size of baseballs, of soccer balls, or of marbles; if it is relatively round and all the same size, it will have about 47.7% voids.

Photo 9. Grinding media. These ceramic spheres illustrate that there is space between spherical particles of the same size.

Quality Concrete from Crap

Consider these round balls. They are ceramic grinding media. No matter what they are placed in, no matter how tightly they are packed, they will always have 47.7% void space. If they were to be used for concrete aggregate, they would still have to have 47.7% cement paste to fill the voids. That means that 18 bags of Portland cement wold be needed per cubic meter of concrete to get good strength. To get workability in the concrete, a little more cement paste would have to be used. Since Portland cement has a higher cost than most aggregate, that would increase the cost of the concrete.

To make strong concrete with uniformly-sized gravel, use as much cement paste as gravel. If less paste is used, the concrete will be weak and have holes in it.

Photo 10. Spheres of different sizes. A bucket of baseballs could hold many ceramic grinding media slipped between the baseballs.

Mix Design Considerations

Since gravel is cheaper than Portland cement, modify the gravel mix so there are fewer voids. Then less Portland cement is needed.

A container of baseball-sized round aggregate would still have the same 47.7% void space. With a container of baseball-sized aggregate, a lot of the small round aggregate could be fitted in between the large aggregate.

With perfectly round aggregate, the densest aggregate mix could be achieved by using the Rule of 4. Start with 4 cm aggregate. Fill the void spaces with 1 cm aggregate. Fill the remaining void spaces with 0.25 cm aggregate.

> *Author's Note: In 1958, as a 17-year-old who knew everything, using a slide rule (a device used before calculators were available), I devised the Rule of 4, and then found that if I had read the literature, it was common knowledge. Lesson learned—Learn from what others have done.*

Ideally, the small gravel should have a diameter ¼ the diameter of the large gravel, and the sand should have a diameter of ¼ the diameter of the small gravel. IDEAL never is found on a construction site. Make the best of what is available.

To obtain the strongest concrete with the least amount of Portland cement, devise an aggregate mix so that it is the densest mix possible. The following section shows a way to do this.

Getting the Most out of Aggregate That Is Available

Aggregate needs to be graded (an aggregate is "graded" if it contains a variety of particle sizes, from large to small), so there are as few pore spaces as possible and as few fines as possible that are less than 100 mesh in diameter.

It is possible to mix large aggregate and smaller aggregate and get a denser mix. To get the densest mix possible, it is necessary to do testing. Here is the process. Use 1-liter containers, 21-liter buckets, or any other container, if the largest aggregate is not more than 1/3 the diameter of the container.

The densest aggregate blend can be obtained if each size of aggregate is about ¼ the size of the next larger aggregate. For example:

4 cm rock,

1 cm pea gravel,

0.25 cm sand, and . . .

. . . It is better if all the aggregate has the same moisture content as it will have when the mixing of the concrete starts.

Repeat the testing with each of the sizes of aggregate. Fill each container up to the same level; or for more accurate measurements, fill the containers level-full of the different aggregates.

Measure the amount of water added to cover the aggregate. With a 21-liter bucket that is level-full and the aggregate is relatively round and all of one size, approximately 10.5 liters of water will be needed. Since the aggregate will not be relatively round and all the same size, probably less than 10.5 liters of water will be needed.

Record the amount of water used to fill the spaces between each of the aggregates used.

Start with a measured amount of the largest aggregate.

Add an amount of the next largest aggregate equal to the volume of water used to cover the largest aggregate.

Mix Design Considerations

Take the volume of water that would fit in the pores of the second-sized aggregate and add that volume of the third-sized aggregate.

If the aggregate sizes follow the Rule of 4, the volume of the blend of aggregates will be the same as the volume of the largest aggregate used. In the example, we used 32.76 liters of aggregate, and it took up only 21 liters of space. This would result in an aggregate blend that is 56% greater in density than the density of the largest aggregate that was used. The reason it does not take up 32.76 liters of space is that the finer particles are fitting between the coarser particles.

If everything worked right, but it never does, all the aggregate blend would fit into a 21-liter bucket, and there would only be 1.34 liters of pore space remaining.

Test different aggregates and develop a blend that is as dense as possible.

To make concrete, the pore space needs to be filled with cement paste. The lower the pore space, the less cement paste that is needed. Besides having enough cement paste to fill the pore space, enough cement paste is needed to hold the pieces of aggregate apart slightly. More on this subject later.

Types of Aggregate

Stones can be divided into sedimentary, metamorphic, and igneous.

Sedimentary rock are formed from material settling in water, or from the air, or from being precipitated chemically from solution. In time, they solidify enough to form rocks. Limestone is the most common sedimentary rock found in Haiti.

Metamorphic rock are usually sedimentary rock that has been changed by heat and/or by pressure. The marble in Haiti started as limestone and then gradually changed to marble when exposed to both heat and pressure.

Igneous rock were at one time molten. Besides the difference in the chemical makeup of the magma (lava), the speed of cooling has a great impact on the characteristics of the rock. Basalt is formed when a siliceous magma cools rapidly and becomes amorphous (without crystals); granite is formed if the magma cools slowly and numerous crystals are formed. Basalt and related rock are very common in Haiti. If the magma contains large amounts of dissolved gasses and cools fast, pumice is formed.

Testing Aggregate

Haiti has lots of volcanic rock and lots of limestone. Both can make good concrete, and both can make very poor concrete. First, we will talk about limestone aggregate. Some limestone is hard, while other limestone can be broken by hand. Since concrete cannot be any stronger than the weakest aggregate, we need to eliminate the soft limestone. A second reason we need to eliminate the soft limestone is because when it is run through a crusher, large amounts of ultrafines are produced. This lowers the strength of the concrete that is produced from it.

How to tell soft aggregate from hard aggregate?

Here is an easy way. Place the aggregate on a large hard stone and hit the aggregate with a sledge hammer. If dust is formed, it is soft aggregate.

Soft aggregate can be broken by grinding it with a sledge hammer. Soft aggregate can be broken by tapping it softly with the sledge hammer. When soft aggregate is broken, ultrafines are usually produced.

Hard aggregate, when it breaks, usually breaks with a smooth-shiny-surface and sharp edges.

Mix Design Considerations

Larger pieces of hard aggregate, when knocked together, produce a ringing sound. Soft aggregate is more likely to produce a thud. If there is an internal fracture in the hard aggregate, it might also produce a thud.

After one gets familiar with hard and soft aggregate, the relative hardness can be determined by looking at it.

When looking at hard aggregate, there are often sharp edges; or if water-worn, the stones are rounded. Soft aggregate tends to look more eroded.

Photo 11.   Hard limestone aggregate.   Hard aggregate breaks with sharp corners and very little ultrafines. Hard aggregate cannot be broken by hand and usually requires a sharp blow of a hammer to break it. Normally, grinding with a 0.45 kg hammer will not break hard aggregate.

Photo 12. Soft limestone aggregate. Rubbing on it with a thumb may produce dust. Soft aggregate can usually be broken with a light blow from a hammer. Normally, grinding with a 0.45 kg hammer will break soft aggregate. When soft aggregate is broken, some fine dust is usually produced.

Photo 13. Unsound limestone. Some aggregates deteriorate when exposed to environmental conditions for a few years.

Sizing Aggregate

Author's Note: Over 50 years ago, I spent some time at a southeast Asia aggregate plant which was producing aggregate for the US military. There were piles of aggregate, and there were men, each with a round ring, and some with hammers. They crushed and sized the aggregate.

Here is a primitive way to size aggregate.

The first pile of aggregate came from the quarry, which we will label *Pile No. 1*. There were stone which were 30 cm in diameter. Most of it was smaller. A portion of it was dust. A man would pitch any piece which would pass through his ring to *Pile No. 2*. Any piece which would not pass through, he would hit with the hammer until it was in pieces which would pass through a 5-cm ring. He would then pitch them to *Pile No. 2*.

Sizing Aggregate

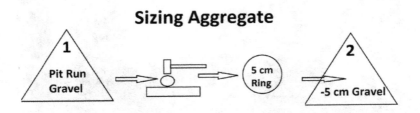

Figure 3. Sizing aggregate. First step.

On the other side of *Pile No. 2* was a man with a 2.5 cm ring. He did not use a hammer. Any piece which would not pass through that ring, he would pitch to *Pile No. 3*. It was for aggregate which was 2.5 cm to 5 cm in diameter. Periodically the aggregate in *Pile No. 3* was shoveled into a wheelbarrow and hauled away.

Sizing Aggregate

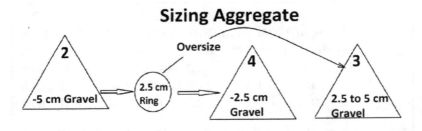

Figure 4. Sizing aggregate. Second step.

The rest of the aggregate went to *Pile No. 4*. On the far side of *Pile No. 4* was a man with a sieve with 0.6 cm holes. Aggregate which would be caught on top

Mix Design Considerations

of the sieve was 0.6 cm to 2.5 cm aggregate, was thrown into *Pile No. 5,* and was periodically shoveled into a wheelbarrow and hauled away. The aggregate which would pass through the sieve went to *Pile No. 6.*

Aggregate from *Pile No. 6* was shoveled onto an inclined screen, and the fines which passed through the screen were thrown away. Any material which slid down the inclined screen was placed in *Pile No. 7.* This was sand and was periodically shoveled into a wheelbarrow and hauled away.

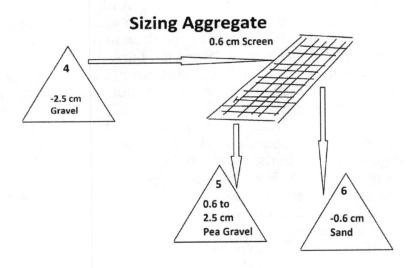

Figure 5.   Sizing aggregate.   Third step.

A modern sand and gravel plant is a mechanized system to do the same thing. Since such a plant is expensive, the following paragraphs will discuss a system which is much more efficient than the ring system, but much less expensive than the modern sand and gravel plant.

The basic unit is an inclined screen. It can be one meter wide and two meters long. It should have more than a 50-degree angle (from horizontal) and less than a 60-degree angle. The concept is to allow the aggregate to slide down the inclined screen at a rate of speed so that undersized material can fall through the holes. The frame can be built from angle iron or from No. 3 rebar or other materials. For a temporary system, a wood frame can be built. The legs need to be stronger than No. 3 rebar.

After the frame is in place, horizontal bars should be placed and welded across the frame. No. 3 (1-cm) rebar will work. No. 3 smooth bar, if available, would be even better. For maximum size 5 cm aggregate, the space between the horizontal bars should be 5 cm apart. That is the space between the bars, not the center to center distance. For maximum size 2.5 cm aggregate, the space between the horizontal bars should be 2.5 cm apart.

After the horizontal bars are in place, then vertical bars need to be placed. Ideally, they should be 1.9-cm angle iron, with the angle pointed away from the frame (upward). For the 5-cm aggregate, the space between the vertical bars should be 5 cm apart. The angle iron will allow all oversized material to slide down the screen.

If angle iron is too expensive or not available, then No. 3 smooth bar or rebar can be used.

For the 0.6 cm screen, build one like the 2.5 cm screen, but instead of adding the vertical bars, add 0.6 cm wire mesh.

For the sand screen, build a 0.6 cm screen and then top coat it with screen wire for windows. It will need to be replaced periodically and is slightly on the coarse side, but it is an economical alternative to trying to import screening which would normally be used in a modern sand and gravel plant. If the fines are wet, they will not go through the screen unless they

Mix Design Considerations

are washed through the screen, thus it is better if the fines are dry. If they are not dry, a system needs to be added so water can be applied to the screen to wash the clay through the screen.

With the fines which are collected, do not use them for concrete, mortar, or stucco. They may be used for fill.

Impact of Water/Cement Ratio on Strength

One kilogram of Portland cement needs about 0.35 kilograms of water to fully hydrate. Such a mix tends to be stiff and not workable. More water is added to make the paste more workable. More water makes the paste weaker. Abram's Law expresses the impact of water on the strength of concrete and is expressed as follows:

Abram's Law

$$S = K \left[c/(c + w)\right]2$$

S = Compressive Strength

K = Constant derived from components used

c = absolute volume of Portland cement

w = absolute volume of water

The formula was developed so that if a lab tech tested one mix, an engineer could then calculate the strength of mixes with differing amounts of water and cement. In deriving the "K," it can predict the compressive strength in mega pascals, pounds per square inch, or as a percentage of the original test results. In the classroom, it is common to use the percentage of the original test results.

For this exercise, the absolute volume of Portland cement is one cubic foot (28.3 l) (1 bag). It takes 32.9 pounds (14.95 kg) of water to hydrate a bag of Portland cement. That is 0.53 cubic feet (15 liters). Fix S at 100 (as in 100% strength) and solve for K. The answer is that K is 234.

Now comes the fun part. Assume that the amount of water is doubled. The formula becomes:

$$100S = 234 \left[1/ (1 + 1.06)\right]2$$

$$100S = 234 \left[0.236\right]$$

$$S = 55$$

The strength of the mix will be 55% of the strength of a mix where the minimum amount of water was used. See how easy it is to weaken concrete? Put in enough water to get the workability needed, but not one drop more.

After samples are produced, cure them and break them in a press to obtain the compressive strength. Compressive strength is usually measured in megapascals (MPa) or in pounds per square inch (psi). For example, concrete which would be used for a sidewalk might have a compressive strength of 17.25 MPa or 2,500 psi. Concrete which is to be used for a house slab might have a compressive strength of 27.6 MPa or 4,000 psi.

Knowing the absolute volume of water and Portland cement used, after a K factor for that mix is developed, estimates of the strength of the mix with varying amounts of water and Portland cement can be made.

Impact of Air Content on Strength

When mixing concrete by hand, do not worry about adding air content to the mix. When upgrading to mechanical mixing equipment, it is easy to add air content to the concrete. Air is often added to concrete where freezing weather occurs, to keep the freezing and thawing from breaking the concrete. In Haiti freezing of concrete is not an issue. There are companies which make water-reducing agents, so a better water/cement ratio can be maintained. Some, but not all, of those agents are soap, and they add air.

Ferret's Law treats the absolute value of entrained air in the same manner Abram's Law treats the absolute volume of water. Ferret's Law was developed first, and Abram's Law is a simplification of Ferret's Law.

Ferret's Law is written in this manner:

Ferret's Law

$$S = K [c/ (c + w + a)]2$$

S = Compressive Strength

K = Constant derived from components used

a = absolute volume of air

c = absolute volume of Portland cement

w = absolute volume of water

Many plasticizers or workability agents that are sold are air-entraining agents. They increase the workability of concrete.

Many water-reducing agents which are sold are air-entraining agents. They do reduce the amount of water used, and the operator thinks the strength of the concrete is being enhanced, but it is not.

If concrete contains more than a few percentage points of entrained air, problems besides reduced strength can result. Water can wick up the concrete due to capillary action. The height of the wicking is determined by the air particle size and the connectivity of the air particles. That means lots of tiny bubbles increase the amount of water that can wick up a wall. This water can then evaporate from the surface of the concrete, leaving chemicals behind which were dissolved in the water. If the water contains chlorides, the chloride concentration in contact with steel rebar is enhanced; and the corrosion of the rebar, with the resulting expansion of the rebar, results in the concrete cracking. See the section on *Chlorides and Concrete* and *Addendum B* for a full explanation.

If concrete is not consolidated by rodding (moving a rod up and down in a column or in a massive pour to eliminate air pockets and ensure that all voids are filled), jitterbugging (mechanically vibrating the top of a slab to ensure that all voids are filled), or other means, there will be large air pockets. These do not reduce the strength of concrete in the same manner as the entrained air does. They have a <u>stronger</u> impact on the strength, but I have not figured out how to reduce it to a mathematical formula, YET.

The big impact on strength comes when there is not enough cement paste to fill all the voids between the gravel particles. This results in an absolute volume of air being up to 3 times the absolute volume of Portland cement. These situations reduce the strength in the same manner as the air voids from lack of rodding or jitterbugging.

The following example calculates the impact on the strength of concrete if there is not enough cement

Mix Design Considerations

paste to fill the voids between the aggregate particles, which results in 0.5 cubic feet of air space (for this example, assume that these voids are only as bad as entrained air; they are worse), and that twice the minimum amount of water was used.

$$100S = 234 [1/ (1 + 1.06 + 0.5)]2$$

$$100S = 234 [1/ (2.56)]2$$

$$100S = 234 [0.153]$$

$$S = 35$$

The strength of the mix will be 35% of the strength of a mix where the minimum amount of water was used and the gaps between the aggregate particles were filled.

How well would this concrete hold up for a roof when the next earthquake occurs? Who would be willing to sleep under such a roof if they knew how it was built?

Impact of Non-Reactive Ultrafines on Strength

Some sand has fines in it. If the sand and water are stirred up, after a minute or two the sand settles out, and the ultrafines turn the water cloudy or muddy. By letting the mix sit for several hours, a determination can be made as to which portion of the sample is sand and which is ultrafines. These non-reactive ultrafines impact the strength in the same manner that excess water or excess air does.

There are fines, such as pozzolans, which react with the Portland cement to form more cement hydrates. These enhance the strength of the concrete.

> *Author's Note:* Back in the 1960s my father deter-
> mined that adding more pozzolans to a mix
> than there were receptors for the pozzolan re-
> duced the strength of concrete. I determined
> that when there were non-pozzolanic ultra-
> fines, the same result occurred. So, I carried
> that testing on through the first decade of the
> 21st century and developed a modification of
> Ferret's Law which addresses this problem.

Nordmeyer's Law can be written in this manner:

Nordmeyer's Law

$$S = K [c/ (c + w + a + d)]2$$

S = Compressive Strength

K = Constant derived from components used

a = absolute volume of air

c = absolute volume of Portland cement

d = absolute volume of fine dust

w = absolute volume of water

Some of the sands that are used for mortar and for plaster in Haiti have over 30% clay. They help the uncured mortar to stick, but they reduce the compressive strength of the mortar or stucco. Much of the aggregate used in Haiti is soft limestone. When it is crushed, much of it becomes ultrafines and reduces the strength of the concrete. Nordmeyer's Law ad-

Mix Design Considerations

dresses particles which are less than 200 mesh. If a water slurry of aggregate is mixed, the particles which have not settled within about 2 minutes are the ultrafines. Particles which are a little larger also decrease the ultimate strength of the concrete, but not as much as the ultrafines.

It is easy to estimate the amount of ultrafines in an aggregate mix. Figure 6 illustrates using a clear cylinder to run a soil texture test or a soil profile test. The cylinder is filled about 2/3rds with the aggregate, and then the aggregate is covered with water and the mix is shaken. Immediately, the pea gravel and sand will settle out. Then the fine sand. Following that will be the silt. The water will remain cloudy until all the ultrafines (clay) have settled. This might take up to 24 hours, but a reasonable estimate can be made in about 10 or 20 minutes.

Use a ruler and calculate the percentage of each grade of material in the sample illustrated in Figure 6.

Soil Texture Test

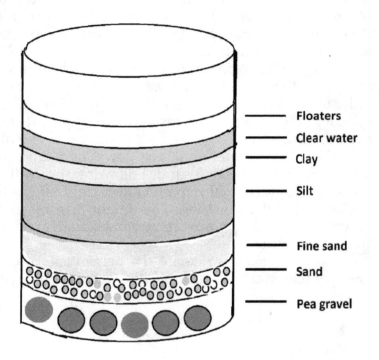

*Figure 6.   Soil texture test.   Fill the container 2/3rds
full of soil; cover with excess water. Shake and
let the particles settle.*

If an ideal cylinder is not available, any clear glass
or plastic container with reasonably straight sides will
do.

Sometimes other things show up. Often, organic
matter will float to the surface. Sometimes, there is oil
mixed with the sand, and some of it coats the surface
of the sand. Oil acts as a bond breaker, so if there is

Mix Design Considerations

any oil on the surface of the aggregate, the Portland cement paste cannot bond to the aggregate.

Also, some people add clay to the concrete mix because it adds volume to the cement paste. Clay is added to mortar and stucco to give them body and to allow them to stick to a wall. Most clay and other ultrafines do not react to build strength. They act like adding extra air or water.

If the sand had 10% ultrafines, and the formula called for using twice as much sand as Portland cement, then the mix would be adding 0.2 units of absolute volume of ultrafines.

With clay in the mix, several additional problems can develop. Over the years, the clay can be washed out of the concrete, stucco, or mortar, and that weakens it even more. Photo 2 (Chapter 3) illustrates what can happen in five years if excessive clay is used. Another problem is that many types of clay tend to hold moisture. If water wicks up a wall, the clay gets wet and keeps the surface of the rebar damp. This leads to corrosion of the rebar. If chlorides are present, this accelerates the corrosive actions. Again, see the section on *Chlorides and Concrete* and *Addendum B* for more details about the corrosive action on rebar.

Ferret's Law and Nordmeyer's Law operate in the same manner as Abram's Law. Rather than remember the three formulae, it is easier to remember Nordmeyer's Law, not because of the name, but because when working with mix designs, all the components usually end up needing to be considered.

Pozzolans and Other Additives

Pozzolans
A pozzolan does not have any cementing ability; but when in the presence of calcium ions and water, it

forms cement chemical hydrates. When Portland cement hydrates, it liberates calcium ions. If a pozzolan is present, additional cement paste is formed, and the concrete becomes stronger.

2,000 years ago, the Romans used volcanic ash to make concrete. They ground it and mixed it with lime putty and aggregate. They made concrete, mortar, and stucco. It took a long time to cure, but some of it remains in good condition after 2,000 years.

In the 1950s and 1960s, Pozzolana, Inc., produced a natural pozzolan from volcanic ash. This product was used to replace 20% of the cement in the concrete in Falcon Dam on the Rio Grande River in South Texas. It was used in the addition to the Galveston Sea Wall and was used in a dam in the Dominican Republic. In 1957 Péligre Dam was completed in Haiti. It used 13,500 tons of our Rio Grande Pozzolan and is the highest dam in Haiti.

By the late 1960s, fly ash started replacing natural pozzolans because it was cheaper. Fly ash, a byproduct of burning coal, also contains pozzolanic properties. Now, most of the concrete produced in the US contains fly ash.

Metakaolin is clay that has been heat-treated to remove some of the chemically combined water. It also has pozzolanic properties.

Depending on economics, it may make sense to build a fly ash terminal and import fly ash, or it may make sense to investigate the availability of volcanic ash deposits in Haiti and develop a processing plant in Haiti to produce natural pozzolans. The natural pozzolan could be used to replace a portion of the cement used to make concrete. This would lower the cost of the concrete and would make the concrete stronger and more impervious.

Mix Design Considerations

Silica fume is also a pozzolan, but requires considerable testing and water-reducing agents to be used effectively. It is beyond the scope of this book.

Since much of Haiti is volcanic in nature, chances are there are materials which could be used to produce a natural pozzolan. Basalt rock could be used, but the crushing of it would be expensive. Volcanic ash deposits are probably present.

Sugar Cane Bagasse Ash

Sugar cane is crushed to extract the juice. The residue is a dry and pulpy material. When the bagasse is burned (often it is used as a biofuel), the ash residue can be used as an additive to cementitious materials. Since the bagasse contains silica, if it is burned at a low-enough temperature to not drive the combined water from the silica, it makes an excellent pozzolan. Since the sugar cane also contains some calcium, the silica particles and the calcium will react if the ash is not kept dry, killing the beneficial effects. Depending on the burning temperature, some fibers may also remain, which can increase the crack resistance of the concrete or stucco.

Rice Hull Ash

Rice hulls contain between 2% and 4% silica. In the burning of the rice hulls, the ash will contain a silica hydrate which is pozzolanic if the temperature is kept low enough. Plants in the US which burn rice hulls seldom collect it for use as a pozzolan, but use it for its high absorbency. To be used as a pozzolan, the rice hulls must be crushed after burning.

> *Author's Note: In the 1980s, working with rice hulls and crushed limestone, my father and I developed a process whereby a low-grade cement which would be adequate for mortar or stucco could be produced. It was part of a larger rice hull processing research proposal. The company took our proposal and turned it over to a university to research. They and the university were not interested in village-scale cement production.*

Plasticizers for Mortar and Stucco Formulae

Traditionally, hydrated lime was used as the cement in mortars and stuccos. When Portland cement came on the market, small amounts were added to speed the set of the lime mortar or plaster. With time, the amount of Portland cement increased. Then it became common to add hydrated lime to Portland cement to get the preferred workability.

In the 1920s, masonry cement was developed. Traditionally, it was produced by inter-grinding limestone with Portland cement clinker and Vinsol resin. Mixing the three components without the inter-grinding does not produce as good a mix.

There are hundreds of products referred to as mortar fat. Two of the most popular are Kel-Crete and Easy-Spred.

Kel-Crete is made from a derivative of guar beans.

Author's Note: I plan to test Guar Beans to see if simple processing can make them usable as a mortar and stucco addition. If they can be used, this might be a crop that Haitian farmers could grow to add to their income.

Easy-Spred is reputably made from diatomaceous earth and sodium bentonite. The diatomaceous earth is amorphous silica and is a pozzolan. The sodium bentonite reacts with the calcium ions released when cement hydrates and forms calcium bentonite, which adds strength to the mortar.

Author's Note: In the 1940s E. H. Nordmeyer sometimes burned the spines off prickly pear cactus, diced it, crushed it, mixed it with water, and let it soak for two days. He added it to a Portland cement/sand mix for making mortar and stucco. He always said that he did not make up more than was needed because after two days it started stinking. It gave body to stucco and reinforced it with the nopal fibers.

Hydrated Lime

Currently, hydrated lime is not being imported into Haiti. It is used for many purposes, including giving mortar body so it will hold masonry units where they are placed, improving the bond between the mortar and the masonry units, reducing water penetration through masonry joints, and improving the workability of stucco so it will stay on the wall or ceiling where it is placed. It is also used for whitewashing of masonry and concrete structures. This reflects light and prevents the sun from heating the building as much, and it protects the concrete surface.

Steps need to be taken to start importing hydrated lime into Haiti and training masons and plasterers to use it.

Depending on economics, it may make sense to investigate the availability of limestone rock deposits in Haiti and to develop a processing plant in Haiti to produce quick lime which can then be hydrated into either hydrated lime or into lime putty. Such a product could be used to produce masonry cement, plasters (stucco), and even concrete, if care is taken.

A processing plant could be as simple as a pit dug in a hillside and fired from below with wood to produce quick lime. Then the quick lime is slaked (mixed with water) and buried in a trench about 0.5 meters wide and a meter deep, so it can cure for 3 to 6 months to produce a hydrated lime paste.

On the other extreme would be a lime plant consisting of a rock crusher, a rotary kiln fired with natural gas, fuel oil, or powdered coal, and a pressure hydrator to produce powdered hydrated lime. This would be followed by silo storage or a bagging plant to bag the hydrated lime powder.

Author's Note: In the mid-1940s E. H. Nordmeyer burned limestone, hydrated it, and mixed it with a Portland cement/sand mix. He would never let anyone get close, because during the hydration process, the burned rocks would sometimes explode.

Mix Design Considerations

Mortar Fat

Mortar fat is a name that is applied to many different products which tend to do the same thing to stucco and mortar that hydrated lime does. They are also called plasticizing agents and are used in concrete to improve the workability.

One of the more common ones is called Kel-Crete. It is made of a derivative of guar beans. With Kel-Crete, only 30 to 45 grams are needed per bag of Portland cement. With hydrated lime, 5 to 10 kilograms are needed per bag of Portland cement.

Currently, most guar is grown in India. Since Kel-Crete is made from guar gum, it is possible that guar beans could be grown and crushed and used as a mortar fat. If the concept works, then work needs to be done on the best way to process the guar beans by hand and to determine the best dosage rate. If it works, then there is the potential of a small industry to produce guar powder to go into stucco and mortar. An added benefit is that guar is a legume and fixes nitrogen from the air into the soil. It could be used by farmers in crop rotation to improve the soil.

Water-Reducing Agents

If the excess mix water in concrete is eliminated, the cured concrete is stronger. As a result, a water-reducing admixture industry has developed. Some agents replace the water with tiny air bubbles. Feret's Law demonstrates that replacing water with air does not improve the strength of the concrete.

Other water-reducing agents do not entrain air and do improve the strength of the concrete. If an engineer has designed the mix formula, do not change it. If a water-reducing agent has been specified, ensure that it is one that does not entrain air.

Accelerators and Retarders

Concrete can become stiff because water has been lost from the cement paste. This may be because of the sun heating the surface and water evaporating, wind blowing on the surface, aggregate which slowly absorbs water, or a substrate which sucks water out of the concrete. In each of these cases, there may not be enough water remaining to fully hydrate the cement particles in the concrete. This results in concrete that is weaker than expected. It may end up powdering or crumbling.

A false set is when non-cementitious reactions, often involving gypsum, cause the mix to appear to set, but no actual chemical hydration of the cement particles has taken place.

Concrete can get stiff because a chemical reaction occurs which causes water to combine with the Portland cement. Accelerators and retarders are designed to impact the speed of the chemical reactions.

For years in cool climates, calcium chloride was added to concretes in winter to accelerate the set. See more on calcium chloride in the next section. Then it was found that the chloride content accelerated the oxidation of rebar, so in most countries it has been banned from use in concrete.

The simplest accelerator is warm water. The simplest retarder is cold water.

Where there is a need, there are cements which will set in 2 hours and develop normal 28-day strengths within a day or two. Many, but not these cements, produce a great deal of heat when curing, and care must be taken when using them to prevent thermal cracking from occurring.

Getting into these specialty cements is beyond the scope of this book.

Chlorides and Concrete

For years calcium chloride was added to concrete to accelerate the set, especially in cold regions. Then it was found that the chloride ion caused a deterioration of any steel in the concrete. ASTM standards are now limiting the use of chlorides in concrete, stucco, and mortar. Admixtures cannot be used if they add more than 65 ppm soluble chlorides.

How much is 65 ppm soluble chlorides?

To 1 metric ton of concrete

can add 0.065 kilos of chloride (65 grams).

Each liter of sea water contains 20 grams of soluble chlorides.

Therefore, could add 3.25 liters of sea water without violating that standard.

Each metric ton of concrete contains about 160 kilos of cement.

A metric ton of concrete requires about 80 kilos of water (more is often used).

3.25 liters of sea water per 80 liters of water is all the sea water that can be put in a metric ton of concrete without violating the standard and risking the degradation of the steel.

Most of the water in the lowlands of Haiti is brackish or slightly brackish. That means it has chlorides in it. This can cause deterioration of stucco lath and rebar.

How much chloride content is in the local water? If it is over 800 parts per million of chlorides, there is a risk of causing deterioration of steel. That would be about 1,300 parts per million of <u>salt</u> in the water.

Using the lowest water/cement ratio possible will lower the amount of chlorides added.

Quality Concrete from Crap

Is the chloride content enough to require that rebar be painted before it is installed? If the rebar is to be painted, it needs to be painted with a red oxide paint. Normally, such a paint job will add 5 to 10 years to the life of the rebar. Will that be enough to address the corrosion problem?

Is there a water source which has lower chlorides?

An alternative is to use reinforcing which is impervious to oxidation and other methods of corrosion. Basalt rebar is discussed in Chapter 7.

Brackish water and sea water do not have as much of a negative impact on concrete as they do on the steel that is used in concrete.

Addendum B is a technical report on the corrosion of rebar via oxidation and how chlorides enhance that oxidation.

Formula Notes

Bounce all buckets of Portland cement on the ground several times to get the air out and to compress them, otherwise the amount of Portland cement called for in the mix will be shorted.

Normally, sand is measured in a loose, damp condition. Sand in this condition is slightly fluffed and takes up about 20% more space than sand that is perfectly dry.

If dry sand is used, often the water and cement paste do not wet it adequately, and there are air bubbles that stick to the sand particles. This lowers the strength of the resulting concrete slightly.

If the buckets are not filled brim-full, mark each one so the measurements will be consistent.

Rounded sand and gravel results in more workable mixes. Crushed sand and gravel results in a stronger concrete which is harder to work.

Chapter 5
Mixing Concrete

By Hand

Concrete is often mixed by hand on the ground, in a wheelbarrow, or in a mortar box. The problem with mixing it on the ground is that moisture in the concrete may escape into the ground, and the soil may end up being mixed into the concrete, lowering its strength.

If mixing concrete by hand, dry-mix before adding water, otherwise extra water will be needed to get the entire mix uniform.

> *Author's Note: When I was about 60 and my father watched me mix stucco in a wheelbarrow, he told me that no one under 75 has enough sense to mix stucco. I used 1.5 cm cuts with my hoe, and he insisted that to get a good mix, one should not use more than 1.0 cm cuts with the hoe.*

Use as little water as possible. A little extra mixing may result in a creamer mix, and no extra water is needed.

After the concrete is mixed, use it within 30 minutes. Microscopic bonds start to form within 30 minutes; and if it is moved around after 30 minutes, some of the chemical bonds which have formed will be broken. Once broken, those bonds will not re-form, and the concrete will not be as strong as it would have been if the concrete had not been moved.

> *Author's Note:* *I define stucco, mortar, and surface bonding cement as forms of concrete; therefore, it is appropriate to add formulae for stucco, mortar, and surface bonding cement in the next several chapters.*

Portable Mixers

Portable mixers may be electric-powered, gasoline-powered, or diesel-powered. They can mix concrete more thoroughly than normal hand-mixing. That and labor savings are their advantage. If one is mixing a less-than-ideal concrete, a portable mixer helps one arrive at failure much sooner. There is also the tendency to add extra water to the mixer. This leads to segregation of the aggregate and to lower-strength concrete. The mixer also can grow legs and walk off the job. As with any tool, it can improve performance if used right.

When using a portable mixer, always add most of the water first, about a third of the sand, and then all the fines (Portland cement, pigments, fiber, admixtures); then add the rest of the sand. If more water is needed, add it slowly so the final mix has as little water as possible.

Never believe the manufacturer's rated capacity. Often, they measure the contents of a mixer by how much water it can hold if filled to the brim. Consider that 50% of the manufacturer's rated capacity might be a better estimate for the capacity of the mixer.

The most common portable mixers have a rotating tub that is set at an angle. They are designed for mixing concrete, but with care can be used for mixing mortar or stucco.

67

The horizontal shaft mixers are designed for mixing mortars and stuccos. If they are used for mixing a product with coarse aggregate, they will lock up.

Ready-Mixed Concrete

In the United States, most concrete is delivered to the job site in ready-mix trucks. The most common kind has a rotating drum so the concrete mixes on its way to the job site. Years ago, most of these trucks were in the 4 and 5 cubic meter capacity, but by adding extra axles and improved roads, many are 9 cubic meter capacity. Maneuvering such a truck through some of the streets of Haiti and in other areas where the roads are not well developed would be a problem.

There is also a kind which has hoppers for sand, gravel, and cement. At the job site, the components are volumetrically metered to a mixer, and the concrete is delivered fresh. These trucks are best where small amounts of concrete need to be delivered over several hours. Again, maneuvering such a truck through some of the streets of Haiti would be a problem.

For small jobs, an alternative is a mixer trailer which carries just over a cubic meter of concrete and is towed behind a small truck.

A serious problem with any ready-mixed concrete is that if a poor formula is used, large quantities of poor concrete can quickly be made, and numerous failed jobs can be completed.

Ready-mix trucks need to be filled before delivery of the concrete. This usually requires a plant with a cement silo, an accurate method of weighing the components, and a system for loading the components into the ready-mix truck.

If a ready-mix truck gets stuck, or gets tied up in traffic, and the concrete sets, it is nearly impossible to clean the concrete out and reuse the drum; therefore, the driver needs to have a method of stopping the cement hydration process. The most common method is for the driver to carry a sugar source. Once the sugar is added, the concrete is worthless.

Testing Concrete

No matter how the concrete is mixed, it must contain the minimum amount of water needed to develop strength, but enough water so that it is workable.

Mortars and Stuccos

The easiest test is the trowel test, where the product is placed on a trowel, and the trowel is held at about a 45-degree angle to see if it remains in place. Then the trowel is held at about a 60-degree angle, and the product slowly slides off the trowel. The trowel is then reloaded and flipped, so the product is underneath the trowel. The product should remain hanging there.

Ordinary Concrete

With concrete containing coarse aggregate (foundations, slabs, columns, beams, sidewalks, roads, etc.), a slump test should be performed.

A slump cone is a truncated cone, usually made of metal, but sometimes made from plastic. Each end of the cone is open. The cone is 30 cm high, the lower diameter is 20 cm, and the upper diameter is 10 cm. Handles are usually attached to the exterior of the cone, and flanges are attached so the operator can stand on them and keep the cone from moving. A tamping rod which is 1.6 cm in diameter and between 40 cm and 60 cm in length, with both ends rounded, is used to consolidate the concrete in the slump cone. A measuring device calibrated in 0.5 cm or smaller increments is used to determine the slump of the concrete.

The slump cone must be placed on a flat, level, impervious surface and is filled in three layers. The final layer should extend above the top of the slump cone. As each layer is placed, it is rodded 25 times. After the top layer is rodded, the excess should be struck off with the tamping rod using a screeding and rolling motion.

Carefully lift the slump cone off the concrete and set it beside the concrete cone. Place the tamping rod on the top of the slump cone and extending over the concrete. The slump of that batch of concrete is the distance between the tamping rod and the top of the concrete.

Chapter 6
Foundations & Slabs

Foundations consist of two or more parts. Primarily they are:
- Footings
- Stem wall or plinth

Besides being wherever there is an outside wall, they need to be wherever a loadbearing wall is located in the building. Before discounting the need for internal loadbearing walls, remember, they are often needed to make a building disaster-resistant.

Foundations support the building. If they are not adequate based on the soil conditions where the building is sited and are not designed to handle the loads which the building places on the foundation, extra stress is placed on the building. If the concrete in the building is not designed to handle that stress, and usually it is not, then the concrete in the building fails.

Engineering a foundation can be complex. Hire a competent engineer to design the foundation if possible. If that is not an option and the structure is a one- or two-story building, follow the guidelines that are listed here. If building a three-story building, whether affordable or not, hire a competent engineer.

With building the domes mentioned in this book, they are much lighter than a conventional home, so the foundations do not need to be as massive since they tend to "float" on the soil where they are placed. For them, follow the guidelines listed in *Homes for Jubilee* or in *Kay pou Jubilee*.

Author's Note: *The stucco and block walls were cracking on an unfinished building at a major university. The engineer had gone to the literature and determined the soil type and designed an adequate foundation. The contractor had installed it according to the design. What everyone had missed was that there was a landfill that was not shown in the literature, and money had been saved by not drilling and sampling the soil underneath the foundation. That soil should have been supporting the center of the building, but it was not. The center of the building was sinking.*

Footings

Footing Width

First, find out about the width and depth of footings on neighboring houses. If there are any signs of settling or shifting, build a wider and/or deeper foundation.

Second, consider the width of the footing. The width is based on the soil type:

Hard soil, such as rock and gravel – Minimum 40 cm wide

Clay soils and clay/sand soils – Minimum 50 cm wide

Sandy soils – Minimum 70 cm wide

Footing Depth

Consider the depth of the footing into undisturbed soil. It should be at least 50 cm into undisturbed soil. It should also extend above the soil line to elevate the building above the surrounding soil levels. As a minimum, it should extend at least 30 cm above the finish grade around the building.

When digging the trenches for the footings, there are several things to remember. The footings will carry more weight if the trenches have flat bottoms, so dig them with flat bottoms and squared edges. The footings will carry more weight if the sides of the trenches are dug into the undisturbed soil.

If soil must be dug out and then forms have to be built to hold the footings in place as they are poured and cured, they will never hold as much weight as if they were formed as discussed in the above paragraph. The top of the footings should be level. There are times, such as in sandy soil, when a wider trench than the footing must be dug and then forms have to be built. In other types of soil, it should be avoided. Note that sandy soils require a much wider footing than clay soils. This is part of the reason.

When building on sloped land, it is possible to step the footings. If this is needed, it is better to hire an engineer to help design the stepping.

Author's Note: In the section on sizing the foundations, the size may seem extreme when compared with foundations for houses built in the US and other parts of the world. A different school of thought is used in the US. The foundations are smaller, but they are engineered with carefully placed and tied rebar, and a much-higher-strength concrete is used. The larger foundations described in this book are less expensive to build but function as well. If a professional engineer designed the structure, follow his/her recommendations.

Foundation Formula

The foundation is designed to transition between solid earth and the building. It does not need rebar, and it does not need to be as strong as the concrete higher up in the building. It does need to be protected from erosion.

Foundation Formula

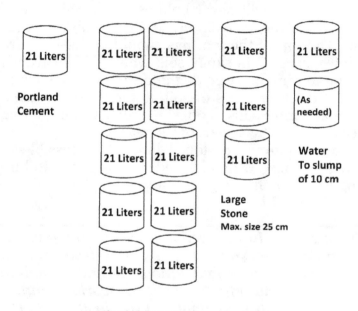

Blended Aggregate

Mix cement, blended aggregate, and water to a slump of 10 cm.

Pour into foundation trenches and consolidate.

Add large stone. Stay away from reinforcement.

Moist cure for 3 days.

Figure 7.   Foundation formula.

The amount of water used will depend on the pore space of the aggregate and the moisture level of the fine sand. Decide on the volume needed to get a slump of about 10 cm.

Large stone can be used, but they will not fit in a mixer, so they are added as the concrete is poured into the foundation and rodded. To obtain a better bond between the concrete and the large stone, all dirt and clay needs to be washed off the large stone before they are added to the foundation.

Keep large stones at least 30 cm away from any rebar which is inserted for a column.

Since the foundation concrete is weak, it needs to be protected. The best way to protect the foundation is to have it located underground. If that is not possible, it should be plastered when the walls are plastered.

> *Author's Note: Back in the 1950s in northern Mexico, many walls were built with a thin sand/cement paste, then large rock were dropped into the forms after the forms were partially full of the paste. By the 1960s, the cement paste eroded on some of these same walls because the cement paste had such a poor water/cement ratio.*

Reinforcing Anchored in the Footings

If confined masonry or infill masonry building technology is used, vertical reinforcing columns are needed. The trusses are anchored about 5 cm above the bottom of the footing and extend up not less than 60 cm beyond the location of the first-floor roof if a second story is envisioned.

With the type of footings discussed here, horizontal reinforcing is not needed. However, at each corner, on each side of each door, and at each side of each

Quality Concrete from Crap

window, a reinforced column needs to be installed. Additionally, if there is any span over 4.5 meters, a reinforced column is needed.

See the section on Confined Masonry for more information on building the rebar trusses to go into the columns and on installing them in the trenches.

Utilities

If the structure is ever going to have utilities coming in underground, it is necessary to install holes through the footings. This is best done by installing PVC pipe a size larger than the lines which will be carrying the utilities, but the maximum diameter of the installed pipe should not be greater than 15 cm. The process varies with when the slab is poured.

If the slab is to be poured later

When installing the PVC pipe, temporarily cap the exterior end and extend the PVC pipe not less than 15 cm on each side of the footings. Then install an appropriate elbow and bring the pipe up to above final height of the base of the backfill. Cap the interior end of the pipe until after the footings and stem walls are poured. Prior to installing the backfill, run appropriate lines and end them at or above the finish grade of the slab. Cap or plug the lines before pouring the slab.

If the slab is to be poured with the footings

If the structure is a dome or if the slab and the footings are poured at the same time, the stub-outs need to be installed from not less than 15 cm outside the footings to their final position. Cap or plug each end.

Backfilling

The area above the original grade of the land and the top of the footings needs to be backfilled. When this is done, the backfill needs to be added in layers and then tamped. While it can be tamped by hand, a mechanical vibrating tamper works better. Tamp the soil in tiers of 2 to 3 inches and be sure it is wetted as it is tamped.

Now is just about the only time when building on sandy soil has an advantage. If the sandy soil is wetted as it is backfilled, it will tend to consolidate and require less tamping. Tamp it anyway.

After the stem wall is poured, additional backfill needs to be added and tamped. The final level should be the anticipated level of the bottom of the slab. To keep from wasting concrete, ensure that the top of the backfill is level.

Stem Wall

The stem wall is poured above the footings and provides a place where the masonry units can be laid. For confined masonry houses, infill masonry houses, SCIP houses, and ICF houses, it should be as wide as the walls and should be a minimum of 30 cm high, except as noted below. It should be formed and poured so the top is flat. With clay soils or sandy soils, the stem wall needs to be horizontally reinforced with trusses.

For SCIP and ICF, the stem walls need to be horizontally reinforced with trusses.

Stem Wall or Plinth Formula

Both terms are used. Some people prefer one term, and some prefer the other. The stem wall is traditionally a wall that connects a footing to the above-ground portion of a building. A plinth is traditionally the base of a column. For this book, we will use the term "stem wall."

Ideally, the stem wall should start below the final grade and extend up to the level of the concrete slab.

Two different formulae are supplied for the stem wall. One is for when the soil supporting the foundation is firm, and the other is for when the soil is either a sandy soil or a clay soil.

Stem Wall Formula for Firm Soils

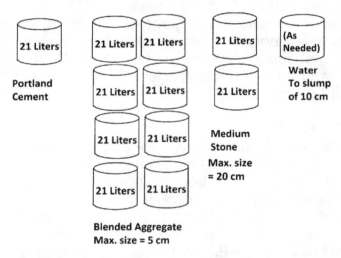

Mix cement, blended aggregate, and water to a slump of 10 cm.

Pour into foundation trenches and consolidate.

Add medium stone. Stay away from reinforcement.

Moist cure for 3 days.

Figure 8.   Stem wall formula for firm soils.

Firm Soil Stem Wall Formula

First, when building on a firm foundation, rebar in the stem wall is not needed; but if rebar is desired, it can be added. If rebar is to be added, the medium-sized stone cannot be used because they would tend to knock the rebar trusses out of line and would interfere with the medium-sized stone passing through the trusses and adequately filling the spaces below the top of the trusses.

Follow the formula in the drawing. As concrete is being poured into the forms which have been set up for the stem wall, medium stone should be added, and the concrete should be rodded. As with the large stone for the foundation, the medium stone should be kept at least 30 cm away from any of the vertical rebar trusses for columns. The width of the stem wall is the width of the wall that will be placed above the stem wall.

Sandy/Clay Soil Stem Wall Formula

For the stem wall for sandy or clay soil, rebar is required. If the area is subject to steel corrosion due to chlorides, a chloride-resistant rebar should be used (such as basalt, fiberglass, bamboo, or stainless steel) and should be built to provide a minimum of 1-inch concrete coverage for any rebar and a wrapping tie every 20 cm.

See the formula on the next page and identified as Figure 9.

Slabs

Slabs can be placed in two different locations. They can be poured inside the stem walls and plinths, or they can be poured on top of them.

Stem Wall Formula
for Sandy or Clay Soils

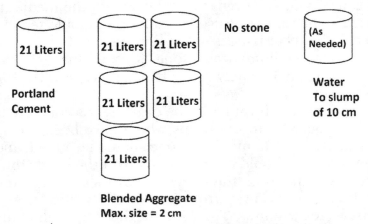

Mix cement, blended aggregate, and water to a slump of 10 cm.
Pour into foundation trenches and consolidate.
Moist cure for 3 days.

Figure 9. Stem wall formula for sandy and clay soils.

Pouring the Slab Inside the Stem Walls

The advantage of pouring the slab inside the stem walls is that no extra forming is needed; and when the slab is poured, it does not interfere with much of the rest of the construction.

Walls can be erected before the backfill that goes under the slab has been completed and compacted.

Pouring the Slab on Top of the Stem Walls

If the slab is to be poured on top of the stem walls, the process is more complicated.

- Arrangements need to be made to anchor the slab to the stem walls.
- Forms must be built and reinforced, if needed.
- After the stem walls have been poured, the forms need to be removed.
- After the concrete has cured, the areas inside the stem walls need to be backfilled and compacted.
- Then forms need to be added to delineate the slab.
- After the slab has been poured and cured and the forms have been removed, the walls can be erected.

Pouring the Slab and the Stem Walls Together

This pour is more complicated than the previous pours and is not recommended for any crew except those with a great deal of experience.

- Arrangements need to be made to anchor the slab to the stem walls.
- Forms must be built to delineate the slab and the exterior of the stem wall.
- Inner forms of the stem wall need to be shorter than the outer forms by the thickness of the slab.
- Backfill and compact up to the top of the inner stem wall.
- Add reinforcing.
- After the stem walls and slab have been poured, the outer forms need to be removed.
- The inner forms cannot be removed and will in time deteriorate or be eaten by termites.

Quality Concrete from Crap

- After the slab has cured, the walls can be erect-ed.

Slab Formula

In many cases the stem wall will be poured with the slab. When this happens, use the slab formula for both the stem wall and slab.

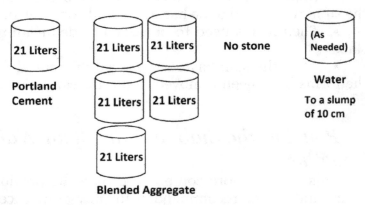

Slab Formula

21 Liters — Portland Cement

21 Liters 21 Liters

21 Liters 21 Liters

21 Liters

Blended Aggregate

No stone

(As Needed) — Water — To a slump of 10 cm

Mix cement, blended aggregate, and water to a slump of 10 cm.
Maximum aggregate size is 1/3rd the thickness of the slab.
If a well-blended aggregate is not available, more cement is needed.
Pour, consolidate, screed, and rough float.
As soon as the water surfaces, fine float.
Moist cure for 3 days.

Figure 10.   Slab formula.   Use blended aggregate as per Chapter 4.

Chapter 7
Reinforcing

Why Reinforce?
Concrete, while strong in compression, is weak in tension. Most movement of cured concrete places some portion of it in tension. Some of the movement is small and driven by chemical reactions that are taking place within the concrete. Examples would be sulfate expansion and reactive aggregate expansion. Such movement normally causes spalling. This is flaking-off of small pieces of the concrete surface. In time, this can cause massive failure.

Other movement can result in rapid structural failure. This chapter is not a design manual for reinforcement that is needed, just an indication where reinforcement is needed and the types of reinforcement which are needed.

Following is a partial list, with brief explanation, of external sources of movement that can cause structural failure in quality concrete:

Footing Failure
While the footings are often not reinforced, reinforcing in the rest of the building may become ineffective if the footings fail.

Footings fail if the ground moves, if the structure above the footings moves, or if the ground below the footings does not have the bearing capacity to hold the weight of the building.

That sounds very simple, but we need to ask, why does the ground move?

If wet clay is squeezed in one's hands, ribbons of clay will ooze out between the fingers. If the clay is only moist, the same thing will happen if more pressure is applied. Remember, the footings are holding the entire weight of the building.

If the soil happens to be clay, or have a high content of clay, a change of the moisture in the soil will often cause it to change in volume. If the soil has been moist and it dries out, it tends to crack.

If there is weight placed on the soil close to the foundation, it may compress some of the soil and put pressure on the footings which may cause the footings to fail.

Most of the readers of this book are in tropical climates. But in northern climates the soil can freeze and expand. Because of this, footings are placed far enough underground where they are not exposed to freezing and thawing.

There are times when the earth moves, such as from an earthquake, and applies stresses first to the footings, and then to the rest of the building. Knowing the likelihood of having an earthquake, and the magnitude of that earthquake in the area where a building is to be located, impacts the size of the footings that are needed.

Every building has weight. The heavier the building is, and the narrower the footings, the greater the pressure which is placed onto the soil that is supporting the building. When this happens, the footing may sink further into the ground. If the entire structure sank evenly, which seldom happens, the only problem would be that underground lines for utilities might be broken, but the building would survive intact.

Reinforcing

A serious problem is that different parts of the building place different loads on the footings. Then moisture getting to the soil around the footings is not uniform all around the building, thus causing reduced load-bearing capacity in different areas. A few centimeters of settling in one corner of a building is enough to put a big crack in a building.

> *Author's Note: I looked at a building recently that had been in place for many years without a problem. When they started landing helicopters on the flat roof, the roof started moving around, compared to the columns that were supporting it. A stress had been added that the building was not designed to handle. When more stress is placed on a portion of a building than that portion of the building is designed to handle, the stress is carried to other parts of the building, and that can cause other portions of a building to fail.*

When wind blows on the building, it causes some portions of the building to push heavier on the footings than when there is no wind. When we are talking about a 1-story or a 2-story building, the stresses are normally minimal. If we are talking about five-story buildings, or more, then the stresses can be much greater. Engineers need to look at anchoring the building through the footings into the ground, so that the building does not turn over.

Quality Concrete from Crap

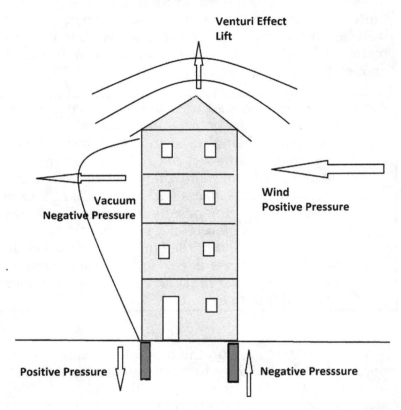

Figure 11. Toppling effect. The higher the building and the higher the wind speed, the greater the forces impacting the building. If the foundation is not adequate, the building can topple.

An important part of reinforcing a building is to ensure that the footing trenches are flat-bottomed, that there are square corners in the trenches, and that the footings are poured on undisturbed soil. This ensures that the weight of the building is evenly disbursed along the soil underneath the footings.

Reinforcing

Stem Wall Failure

Stem walls are a transition between the footings and the building. Usually, any failure of the stem wall is because of movement either above it or below it.

Slab Failure

Slab failures occur for the same reasons that footing failures occur.

In addition to those reasons, weight may be applied to all or a portion of a slab. An example would be when a vehicle drives on it. Or if a heavy piece of machinery is placed on the slab. Or if the heavy piece of machinery vibrates. If the slab is not designed to carry that load, the slab may crack.

Concrete shrinks as it cures. If the ends of a concrete slab are anchored, stresses are placed on the concrete as the concrete shrinks. When those stresses are greater than the tensile strength of the concrete, the concrete will crack. Often, slabs are poured after at least a portion of the walls are in place. Often, there is a piece of wood placed along the walls so the concrete will not bond to the walls. This allows the slabs to float freely, so stresses do not develop in the middle of the slab, which can lead to cracking.

Slabs also fail when there is a source of moisture under the slab. This could be from a spring or from a broken water or sewer pipe.

If the slab is tied to the wall system, any wall movement can crack a slab.

Wall Failure

Usually, walls fail when the footings and stem walls holding them up move, when the footings and stem walls cannot support the weight of the walls, or when shortcuts are taken when the walls are being built.

If the slab is tied to the walls, anything which will crack the slab may crack the walls.

Wind loads can be powerful. Besides blowing against one side of a building, a partial vacuum forms on the other side of the building. Either of these can lead to a wall failure.

During Hurricane Matthew, a downdraft on a roof caused the roof trusses to exert lateral pressure on the walls because there was not a bottom chord on the trusses. Eliminating the bottom chord is fairly common in Haiti.

If a wall is used as a retaining wall, there are added stresses on it; and in many ways, it reacts like a roof does. All retaining walls in homes should be engineered.

Roof Failure

If a wall fails, the stresses are often carried to the roof. This causes the roof to fail.

If the roof fails, the stresses are often carried to the walls. This causes the walls to fail.

With walls, if there is no reinforcement or if there is inadequate reinforcement, the wall may survive without problems for years. Since the lower side of a flat roof is under tension, except where supported by walls, and concrete has low tensile strength, if not adequately designed, it is a disaster waiting to happen.

A concrete roof that is 3 meters by 3 meters and 10 cm thick will weigh approximately 2,250 kilograms. A slab of this size without reinforcing, even if poured from excellent concrete, will probably fail soon after the supports for underside forms are removed. Most roof slabs in underdeveloped countries are not poured from excellent concrete.

If the support of the roof is not completely equal, there will be some areas of the roof which are exposed to greater stresses than others. An example would be an inadequate footing, which might lead to a section of a wall being depressed a centimeter or two.

Many flat roofs are designed with a rim around the edge to trap water. This water, as it evaporates, cools the roof and thus cools the house. A layer of water on the roof that is 2.5 cm thick on a 3-meter by 3-meter roof will add about 235 kilograms to the weight of the roof. Additionally, the water may seep into the concrete and contribute to the oxidation of steel rebar in the roof.

In adding reinforcement to a roof, it is best to add the major reinforcement over the shortest distance. In other words, if a room is 3-meters by 5-meters, the major reinforcement should be spanning the 3-meter direction.

Stresses on a Concrete Panel

If we were to pour a 30 cm (12 inch) thick concrete slab without reinforcement and support it from two opposite edges and test it, we would develop some interesting results. Gravity would apply stress on the panel.

If the temporary supports holding up the bottom form from under the center of the panel were removed three hours after the concrete was poured, the concrete would collapse immediately. If they were removed after 3 days and the panel was not very large, it would probably survive for a while. If it was a large panel, it would probably collapse as soon as the temporary supports were removed.

After the forms and temporary supports have been removed, stress on the lower side next to the supports would be negligible. Away from the supports, the stresses would increase substantially. While a straight edge might not detect a slight curving along the underside of the concrete panel, the panel would be curving slightly. When the stress on any portion of the surface of the panel exceeded the tensile strength of the concrete, a hairline crack would develop. This would weaken the panel; and then at the next point where the stress exceeded the tensile strength of the concrete, another hairline crack would appear.

> *Author's Note: When considering these seemingly harmless hairline cracks, consider the question a doctor once asked me, "How many cholera bacteria do you need to ingest before you come down with cholera?" When I could not answer, he said, "Not as many as you think." It is the same way with those hairline stress cracks on a concrete panel.*

If we applied steel rebar to the sides, then the steel would provide the tensile strength and would carry the load. The amount of stress in the center of the panel would become insignificant. The problem with using rebar to provide tensile strength to a concrete panel is that the steel rebar must be covered with concrete that is at least three times the thickness of the steel. If a 20-cm thick concrete panel is reinforced by Number 4 (1.27 cm diameter) rebar, the rebar must be placed so that it is covered with at least 3.8 cm of concrete. That does not allow the steel to be placed in the areas with the greatest tensile stresses.

Structural Concrete Insulated Panels make use of this concept and will be discussed in Chapter 14 in the section on disaster-resistant structures.

Reinforcing

Placement of Reinforcing

If we wanted to pour a flat concrete roof for a house, we would need to consider where the roof is under tension and where it is under compression. The rebar needs to be in those areas which are under tension.

The underside of the concrete roof is under tension, except where the roof is supported.

The upper side of the concrete roof is under tension above where the roof is supported.

Many times, two continuous layers of rebar are used on a concrete roof—one high in the slab and one low in the slab. This results in more rebar being used than is needed. It does ensure that a failure will not occur between the end of the upper piece of rebar and the beginning of the lower piece of rebar. A compromise is to place continuous rebar close to the bottom of the roof slab; and then above the supports, place pieces of rebar that are at least 1.5 meters long.

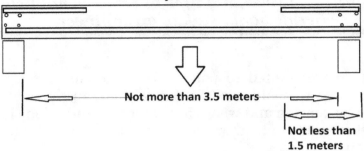

Figure 12. Rebar placement for simple roof. If the rebar moves during the pour, expect failure.

Some people think it is a good idea to place the steel rebar dead center in the slab. That would provide very little reinforcement where it is needed and result in premature roof failure.

Instead of building two walls and placing a roof over them, if two walls were built and a roof was placed over them and hanging over about 2 meters, a different problem would develop. An engineer would be needed to design the reinforcement to handle all the cantilevered weight.

A roof with a 1.0-meter overhang is much simpler.

Rebar Placement
Roof Overhang

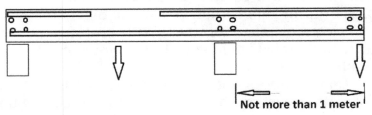

Figure 13. Rebar placement for a simple roof with overhang.

If rebar are placed in the lower portion of that roof rather than in the upper portion of the roof over a support, a roof failure would be likely.

Thus, in designing the rebar placement for a roof, we need to consider where the roof is under tension and place rebar there. This usually results in rebar placed high over walls and other supports and placed low in the spans between walls and other supports.

The steel needs to be protected with concrete. Guidelines vary, depending on the environment and the quality of the concrete. The normal guideline is to cover the steel with an amount of concrete equal to three times the diameter of the rebar. When using 1 cm rebar, cover it with at least 3 cm of concrete.

If basalt rebar is used, the amount of cover need not be as much, since that rebar does not need to be protected from oxidation.

Rebar needs to be lapped and tied where two pieces are joined together. For small-diameter rebar (about 1 cm), lap the pieces about 45 cm. For larger rebar, extend the lap. Use tie wire to tie the two pieces together three times. There should be three ties. Each tie should have the tie wire going around the two pieces of rebar at least 2 times.

Quality Concrete from Crap

When a piece of rebar crosses another piece of rebar, use tie wire to tie the two pieces together. Ideally, with mechanical equipment the rebar mat could be picked up by one corner. An alternative is to weave the pieces of rebar together. With anything heavier than No. 3 rebar, this tends to be hard work. It does result in a strong slab.

Concrete does not bond to rebar that is rusty or dirty. If the concrete does not bond to the rebar, money is being wasted on rebar that is not functioning.

If the design places the rebar in the correct locations, can the people who install the rebar place it in the appropriate locations? If they can, will the rebar stay in the correct locations when the concrete is poured and consolidated? Since the answer is often negative to one or both questions, rebar trusses are often designed using four pieces of rebar; and rebar stirrups are used so that if they are supported high enough above the bottom of the pour, they will provide the needed tensile strength in the appropriate locations.

Deterioration of Rebar

Rebar rusts (corrodes). If it stays damp and has an oxygen source, it rusts faster. If it stays damp, has an oxygen source, and the moisture contains chlorides, it rusts even faster. When steel reacts with oxygen to rust, the rebar increases in diameter. Steel can also react with chlorides in the concrete to form ferrous chlorides; with this reaction, the rebar increases in diameter, but this is minor compared to the oxidation reaction. Any increase in diameter places the concrete under tension, and soon the concrete splits to relieve the tension.

If there are sulfates in the soil and they are wicked into the concrete, the sulfates can react with the steel to form ferrous sulfates.

Once rebar has deteriorated, there are no easy fixes except removing the concrete and starting over.

A common problem in Haiti is that when anticipating an addition to a building, rebar is often left exposed coming out of the concrete roof. The rebar tends to deteriorate faster where it is coming out of the concrete roof. Thus, there is a weak spot where the addition is added. This problem can be solved by cutting the rebar off about 1 meter above the concrete roof and pouring a low-strength concrete around the rebar. When it is time to add the addition, the low-strength concrete can be chipped away.

Alternatives to Steel Rebar

If rebar is subject to attack, there are several ways to protect them. Each method has advantages and disadvantages.

Galvanized Rebar

This is probably the most common factory method of treating steel rebar in a manner so it remains user-friendly. The problem is that the zinc is a sacrificial layer and will deteriorate at some point. Since zinc is highly reactive in an alkaline environment, studies have been performed to determine whether galvanized rebar adds years to the life of rebar. In harsh environments, the results of those studies have been mixed. When galvanized rebar is bent, the zinc coating on the rebar may crack and lower the resistance of the rebar to oxidation.

Painted Rebar

There are several anti-rust paints that can be used on rebar. They can be used on the job site. If they are to be used, the existing rust should be removed as much as possible. Then when they are painted, they need to be placed so they dry without touching anything, so a protective coat can be formed. If two pieces of rebar are touching as the paint dries, when they are separated there will be numerous places where the paint has peeled from the rebar. With paint, the bonding of the concrete is to the paint, and that bond is weaker than the bond between clean steel rebar and good concrete. Often the paint cracks when the rebar is bent. Like galvanization, it will deteriorate at some point, but it does add years to the life of the rebar.

Epoxy-Coated Rebar

They can be epoxy-coated in the factory. This is a more permanent solution, but the bond between the epoxy and the concrete is not as strong as the bond between steel and concrete. Bending the rebar often cracks the epoxy coating.

Cathodic Protection of Rebar

The oxidation reaction which destroys rebar is associated with a very weak electric current. This weak electric current can be neutralized by inducing a direct electric current of the opposite charge into the rebar. This is used only in harsh oxidation environments and is not considered appropriate for use with small structures like houses.

Stainless Steel Rebar

This is a special-purpose rebar and is expensive.

Fiberglass Rebar

Fiberglass rebar, also called glass fiber reinforced polymer (GFRP), is resistant to the oxidation reactions which occur with steel rebar. It is more expensive, and there is a learning curve when using it, since it cannot be bent. Preformed shapes, such as corners, can be ordered and tied to straight pieces of the GFRP.

Basalt Rebar

A new solution is to use rebar made from basalt. Basalt is lava rock. The process starts out like the making of rock wool insulation. It is melted and forced through a tiny orifice. After it is cooled, the similarity to making rock wool insulation ceases.

The strands are coated with epoxy and formed into rebar. The resulting rebar is resistant to any chemical attack which may be found around concrete. It has greater tensile strength than steel rebar. As a result, basalt rebar about half the diameter of steel rebar can replace the specified steel rebar. Additionally, it is extremely lightweight. Basalt rebar for a house slab weights about 11% as much as the steel rebar for a house slab.

Besides making rebar, the strands can be twisted to make rope. This allows for a flexible reinforcement. Since the bond between concrete and the basalt is excellent, and it is resistant to oxidation, the depth of cover can be substantially lower than the depth of cover of steel rebar. As a result, there are situations where much less concrete is needed for a job.

The basalt rope is excellent for reinforcing domes where one can use 5 cm (2 inches) of concrete rather than the 10 cm (4 inches) required when using steel rebar.

Basalt rebar is different from steel rebar. It can be bent into an arch, but as soon as it is released, it returns to its original straight shape. This is referred to as "having no memory." As a result, it can be shipped in coils which are 100 meters long, but are still easy for one man to carry.

This also results in the inability to bend corners and other shapes into the basalt rebar. There are ways to overcome this. Basalt rope can be used to reinforce corners. In using basalt rope to reinforce a corner, keep in mind that per equal diameter, the basalt rope is half as strong as the basalt rebar.

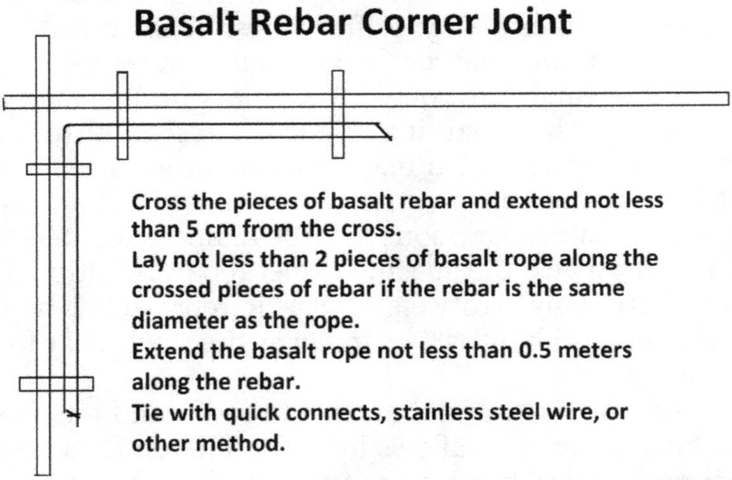

Basalt Rebar Corner Joint

Cross the pieces of basalt rebar and extend not less than 5 cm from the cross.
Lay not less than 2 pieces of basalt rope along the crossed pieces of rebar if the rebar is the same diameter as the rope.
Extend the basalt rope not less than 0.5 meters along the rebar.
Tie with quick connects, stainless steel wire, or other method.

Figure 14. Basalt rebar corner joint. The joint will be stronger if the rope is gently wrapped around the basalt rebar.

Author's Note: My friend Van Smith builds permanent aluminum and galvanized steel scaffolds by wrapping joints with basalt rope and then painting them with epoxy. Such a system could be used with a basalt rebar corner.

Reinforcing

Basalt rope is not to be used as one would use other types of rope. It is permanent only if it is encased in concrete or in a substance like epoxy. Wiggle it back and forth, and the pieces of basalt fiber will come loose; and if there is a unidirectional light source (sunlight), one can often see the fibers separate and float away. Sliding a rope through a hand will embed tiny fibers in the skin. Wear gloves when working with basalt rope, or learn to work safely with basalt rope.

Currently, basalt rebar is more expensive than steel rebar, but the price will come down as it becomes more common. The extra price is often offset by the reduced amount of concrete that is needed. Basalt rebar is not attacked by any of the corrosion problems that beset steel rebar, so it is an excellent choice for a permanent structure.

Bamboo Rebar

Bamboo, if it is processed right, can make an acceptable reinforcing rod. There is limited technical literature available concerning testing and using bamboo as reinforcement.

The bamboo stalks need to be mature. That is three years old. If younger, they may be more likely to be weaker and more likely to rot.

Reinforcing strips need to be dried. Bamboo changes size with moisture changes, so it needs to be as small as possible when placed in the concrete.

The concrete used needs to be as dense as possible to keep the moisture level of the concrete constant and thus the moisture level of the bamboo constant.

Bamboo does not have the tensile strength of steel, so about five times as much cross section of bamboo is needed to replace the cross section of steel in the concrete.

An investigation needs to be done to determine which varieties of bamboo work best for reinforcement of concrete.

Plastic Bottle Rebar

Sometimes someone makes a stupid remark to me, and then several weeks later I think that maybe it was not so stupid.

Currently, we are bringing basalt rope and basalt rebar into Haiti and using it where steel rebar will not hold up. My friend said we should replace the basalt rebar and the basalt rope with plastic bottles.

If a cutter were designed and built which would cut a 1/8-inch strip of plastic bottle, then the strands of plastic could be braided to make a rope. That rope could be used to replace the basalt rope. Would it be strong enough? Add another strip of plastic if it is not. A nice thing about the plastic rope is it would not tend to slip like the basalt rope does. We would need to tie it fewer times.

If we dipped the plastic rope in epoxy, we could make plastic rebar.

Since most plastic used to make water bottles will melt at about 260 degrees C, it might be possible to pass the rope through a heat chamber and allow enough of it to melt to stiffen it. Most plastic when burned gives off toxic fumes, so it would be necessary to experiment to determine whether the process could be done at a temperature where toxic fumes were not given off.

There are questions which need to be answered.
- What would be the economics of making plastic rope and rebar?
- How could we suspend the plastic rebar while the epoxy dried?

- How well would the concrete bond to the plastic rope and rebar?
- How much concrete cover would we need to bond the rope and rebar to the concrete?

Basically, for the cost of a bottle cutter and a simple rope-twisting machine, a family could be in the plastic-rope-making business.

Chapter 8
Beams and Columns

Columns

Columns can be decorative or structural, but often they serve both purposes. Columns can be incorporated into walls of a structure or they can be free-standing. They can be circular in cross section, square, rectangular, or numerous other shapes.

Columns are poured between wall panels of concrete block which have been laid. Boards are placed on each side of the wall panels to keep the concrete in place. Corners are a little bit more complex to form. Ideally, the entire column should be poured at once, but that is not possible when there is a beam installed in the middle of a wall. In those situations, the column form boards should be installed up to the height of the top of the beam. Then the beam form boards should be installed. After that, the beam and the column should be poured at the same time. All the concrete should be well rodded to ensure it is well consolidated.

Since portions of columns are in tension while other portions are in compression, the columns need to be reinforced. The most common way is with rebar trusses built from four pieces of rebar. Periodically, the rebar are held in place with stirrups.

Of course, the rebar trusses interfere with the placement of the concrete, so extra care needs to be taken to consolidate the concrete. At any point on the column where the concrete is not well-consolidated, the steel rebar is apt to corrode. This can lead to the column failing.

Forms for columns need to stay in place until the concrete has cured enough to carry the immediate load that is placed on the column. If no load is placed on the column, then the forms can usually be removed within 24 hours. If the columns are to carry a substantial load, such as a roof, the curing usually needs to last for over a week before that load is placed on them.

Wall Beams

Beams may be described as columns which are lying on their sides. While they can be used for decorative purposes, usually they are structural. The most common wall beams are poured in confined masonry construction or along the tops of walls.

Often, the wall beams are not under tension or under compression unless the wall is under stress. Since stress can come from several directions, it is best to reinforce wall beams with rebar trusses built from four pieces of rebar. Periodically, the rebar are held in place with stirrups.

As with columns, the trusses interfere with the placement of the concrete, so extra care needs to be taken to consolidate the concrete. At any point on the wall beam where the concrete is not well-consolidated, the steel rebar is apt to corrode. This can lead to the beam failing.

If the home does not have a concrete roof, each masonry wall needs to be topped with a concrete beam to stabilize it.

Usually it is safe to remove the forms for wall beams after 24 hours.

Roof Beams

The major difference between wall beams and roof beams is that the roof beams are not supported from below. This means that a portion of each roof beam is constantly under tension. As a result, extra care needs to be taken to ensure that the concrete is well-consolidated and the reinforcement is precisely placed.

Usually, in a concrete roof there are roof beams interspersed with space fillers.

With roof beams, the full load is placed on them as soon as the temporary roof supports are removed. Usually, it is best to leave them in place for at least a week or until testing shows that the concrete has cured enough to carry the load. If the temporary supports are removed before the concrete is adequately cured, the roof might not collapse, but cracks may occur which will lead to failure later.

Column, Wall Beam, and Roof Beam Formula

This formula is like the formulae which are used to pour stem walls when building in sandy or clay-type soils.

Beam & Column Formula

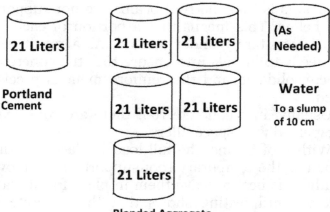

Portland Cement

Blended Aggregate

Water
To a slump
of 10 cm

Mix cement, blended aggregate, & water to a slump of 10 cm.
See Chapter 4 for information concerning blended aggregate.
Maximum aggregate size is 2 cm.
If a well-blended aggregate is not available, more cement is
needed.
Pour, consolidate, screed, and rough float.
Moist cure for 3 days.

Figure 15.   Beam & column formula.

Chapter 9
Concrete

Base Material Under the Concrete
To not use any more concrete than necessary, fill in the low spots and compact it. To avoid thin areas of concrete, cut down the high spots of the base material. If the concrete is going to have vehicle traffic, before pouring the concrete, crushed limestone is often added and compacted in lifts of 5 cm until the desired thickness is reached. This protects the concrete from some of the minor earth movement.

If the cut and fill of base material is not done, be sure and order more cement, sand, and gravel. Remember, compacted base material is much cheaper than concrete.

Forms
Forms should be built and installed before installing the rebar mat. Use something that is handy to build the forms. Lumber, concrete block, whatever is available, can be used. Ideally, they should contain the concrete and have a level surface to screed the concrete smooth. The forms need to be reinforced enough to stay where they are placed.

The forms should come off easily when the concrete has set. While there are lots of form-release chemicals available, used motor oil or used transmission fluid is hard to beat. Each time the forms are used, apply more oil. Each time the forms are removed, scrape any concrete off that is sticking to them.

To keep ground water from wicking up into the concrete, especially ground water containing chlorides and sulfates, place a layer of plastic in the form. Then build the rebar mat. The mat may contain wire mesh as well as rebar.

Pouring

If the pour is too large to do in one day, plan where one day's pour will end and the next day's pour will begin. Cold joints are a fact of life. Cracks often develop where cold joints are located. If there must be a cold joint, add extra reinforcement in the area and slope the concrete so the joint can be wider. Use a concrete bonding agent before pouring the next day.

Before starting the pour, ensure that the rebar mat is in the position it needs to be in. Block it substantially so it does not move as workers walk on it. When pouring a house slab, beam, or column, better results are achieved if it is done in one continuous pour so there are no cold joints.

Consolidation

When concrete is poured, it must be consolidated. As the concrete is being added, it needs to be rodded, vibrated, and/or jitterbugged to ensure that the concrete reaches to the bottom of the forms and is well consolidated. If we were to give human emotions to the concrete, we would say that the concrete loves to separate so the gravel is in a different place than the paste. We would also say that the concrete loves to get caught on rebar so it does not have to fill columns that are there to reinforce the structure.

Photo 14. Unconsolidated stairs. Failure to consolidate this concrete led to a weaker concrete than it should have been. The rebar under the stairway has corroded and expanded. This resulted in the concrete under the rebar cracking and falling off.

The concrete may be consolidated by rodding, starting at the bottom and working upwards. It may be jitterbugged, which is an up-and-down movement on the surface of the concrete. That is not as effective as rodding, but it is much faster if large areas need to be consolidated. A problem with jitterbugging is that it brings fines to the surface, which results in more hairline cracking of the surface. There are power vibrating tools which can be placed in the concrete and quickly bring about consolidation. If they are overused, they can bring about separation of the concrete,

Quality Concrete from Crap

and the problem is then worse than non-consolidated concrete.

After the form is full, screed it off so it is even with the tops of the form boards.

Use a float to remove the screed marks.

Finishing

Back off and wait until the concrete is ready to finish. This is often seen by some of the mix water coming to the surface.

Depending on the use of the concrete, select one of several different finishing techniques. If it is for the floor of a residence, troweling the surface very smooth is a good option. Colored earth or pigments may be dusted on, to add color to the concrete.

If it will be a driveway, a broom finish is a good option. Trowel it smooth and then pull a broom over the surface to add a little texture. Broom finishes can be light or heavy.

Sugar interferes with cement hydration. After a surface is troweled smooth, spray it with Coca-Cola or other sugary drink; the surface cement will be retarded. By the next morning, the deeper concrete will have set, so hose off the surface and have an exposed aggregate floor. Normally, a sugar-and-water solution is cheaper, but the expression on people's faces when it is suggested they spray Coca-Cola on the fresh concrete is priceless.

Keep the surface of the concrete moist for at least 3 days and preferably 7 days. An easy way is to lay burlap or vegetation on the surface after the initial set and keep it wetted.

After the concrete has set, the forms may be removed. This can be done within 4 hours of a pour, but one to three days is better, to prevent chipped corners and edges.

Concrete takes a long time to cure. Conventional wisdom says that ordinary concrete reaches about 90% of its ultimate strength in about 28 days. Conventional wisdom also says that it reaches 75% of its 28-day strength in 7 days. Conventional wisdom often is wrong. Temperature impacts the speed at which concrete cures. Thickness of the concrete impacts how much heat builds up in a pour, since cement particles give off heat when they hydrate. Anytime an additive is added to the concrete, either on purpose or inadvertently, it may change the way the concrete cures.

Chapter 10

Masonry

Foundation

If a stable foundation is not provided under the masonry units, something may move and cause a problem. If the foundation cracks, it will probably crack the wall. If a portion of the foundation sinks, it will cause a diagonal crack across the wall. The width of the foundation wall needs to be based on the anticipated load (weight of the building) and on the load-carrying capacity of the soil under the worst possible conditions. There are clays, that if moist, can be squeezed so streams of clay will be extruded between fingers. Such soil can support a lot of weight when dry, but not support much weight when moist. There are ways to test the soil for its load-carrying capacity and to engineer the foundation needs. See Chapter 6 for a simplified way of sizing foundations.

Mortar

Mortar has two major functions. It holds the masonry units apart, so irregularities in the units do not interfere with building the wall; and it bonds the masonry units together, so the wall becomes, as much as possible, a monolithic unit.

If a Portland cement/sand mortar is used, it does not have enough body to hold the units apart. See the Mortar Formula section of Chapter 10 for mortar formulae.

If mixing by hand, the dry components of a mortar need to be well-mixed before any water is added. Use only enough water to give it the workability that is needed. The mason is the best judge of how much water to use. Among other things, he will judge how the masonry units absorb water from the mortar and end up with a mix which contains enough water for good hydration of the cement particles after the masonry units have sucked some of the water out of the mortar.

With time, the mortar gets stiff. More water can be added, but to prevent loss of bond strength, use the mortar as quickly as possible. Most careful masons prefer that it be used within one hour of the time it is mixed.

The stiffer the mortar becomes, the less mortar can enter the pores of the masonry unit. This substantially reduces the bond strength.

Now we have an interesting problem. Keeping the water/cement ratio low results in higher-strength mortars, but increasing the water/cement ratio usually increases the bond strength. The mason (not the engineer, the building contractor, or the architect) needs to determine how much moisture should be added to the mortar.

Masonry Units

Masonry units include concrete block, brick, and stone. Each has advantages and disadvantages. If the units are good quality, each will function well in a wall.

Much of the concrete block in Haiti tends to be weak. While there are tests, there are few facilities in Haiti which could test the quality of a concrete block. Tests address compressive strength, resistance to erosion, consistent size, and other parameters. A simple test that does not meet any standards is to hold a block at chest height and drop it onto a grass-covered area. If the block breaks, it is poor quality. If it remains in one piece, it is good quality. One of the major problems with this test is that the block are heavy, and there is a tendency to hold the block close to one's chest. This results in the block landing on, or very close to, one's toes.

> *Author's Note: When working with a brick plant in northern Mexico about 30 years ago, the man in charge of quality control told me that a No. 1 brick was in one piece, a No. 2 brick was in two pieces, and that they did not try to sell the No. 3 brick. He laughed and said they had discarded that system several years before.*

Concrete brick should be tested like concrete block.

Fired clay brick should be banged together. If they have a ring to them, they are normally well-fired and will hold up. If they produce a dull sound, they are either cracked or under-fired.

To determine the quality of stone, go to the section on aggregates in Chapter 4.

Mortar Formula

Mortar has several functions. It needs to hold the masonry units apart. It needs to bond the masonry units together. It needs to resist the penetration of water. It needs to stick to the end of a masonry unit as the masonry unit is set into place (small unit) or as another masonry unit is set beside it (large unit). It needs to resist erosion. Many formulae will provide some of the functions, but only a few formulae will perform all the functions.

Ideally, the maximum sand size should be less than 1/3 the width of the desired mortar joint.

The mortar should be rich in cement, so it is easier to tool the joints to produce a water-resistant joint.

A quality mortar will hang onto the trowel when the trowel is inverted, but will easily slide off when the trowel is tapped on a masonry unit.

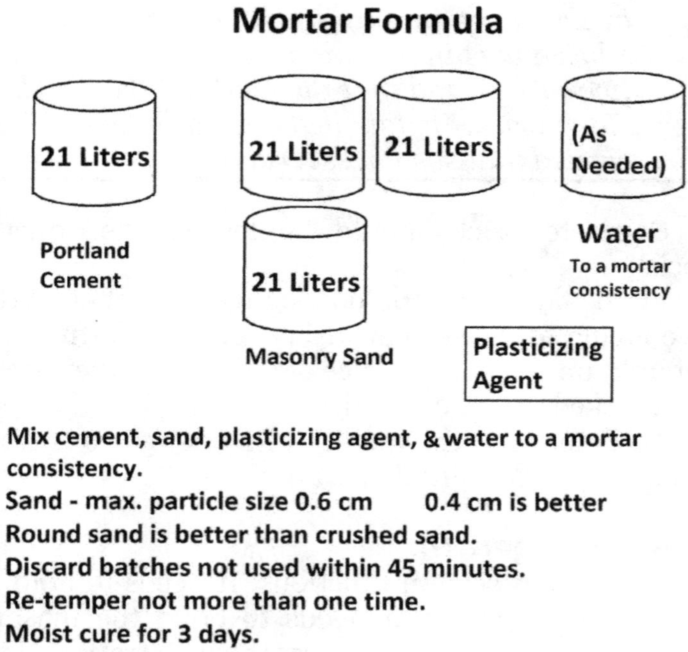

Mortar Formula

21 Liters

Portland Cement

21 Liters | 21 Liters

21 Liters

Masonry Sand

(As Needed)

Water
To a mortar consistency

Plasticizing Agent

Mix cement, sand, plasticizing agent, & water to a mortar consistency.
Sand - max. particle size 0.6 cm   0.4 cm is better
Round sand is better than crushed sand.
Discard batches not used within 45 minutes.
Re-temper not more than one time.
Moist cure for 3 days.

Figure 16. Mortar formula.

In Haiti, and in some other parts of the world, clay is often used as the plasticizing agent. This results in a weaker and more erodible mortar. Often the clay is introduced as part of the unwashed sand that is

Masonry

used. From time to time tests show that up to 30% clay is in the sand that is being used to make mortar. This results in having as much clay as Portland cement in the mix. Other things being equal, the mortar compressive strength will be reduced by about 40%. If clay must be used as a plasticizing agent, use as little as possible.

Plasticizing Agents

Select one of the following to be added to 1 bucket of Portland cement for a mortar or a stucco:

1/4 bucket of hydrated lime, or
20 to 30 grams of KelCrete, or
1.52 kg of Easy-Spred.

Working to develop plasticizers made from:

Guar beans,
Fermented prickly pear cactus.

Figure 17. Plasticizing agents. Determine which agents are locally available and then make the selection. A little extra sand may need to be added with some of the plasticizing agents.

Placing Mortar

If the masonry units are dry when mortar is placed on them, they will suck water from the mortar, and there will not be enough moisture in the mortar for it to hydrate properly and develop strength. If the masonry units are too wet, then the mortar cannot enter the pores of the masonry unit, and an adequate bond will not form. Normally, the block should be well-wetted the day before they are laid.

With smaller masonry units, the units can be hand-held and mortar smeared on them. Concrete block are too heavy to handle this way. The usual practice is to place mortar on the top of the wall and on the webbing of the block which will be receiving the block. Then mortar is buttered on the end of the block to be laid. The block is then placed into position and tamped to ensure it is level and in the correct place.

Laying

If the block must be picked up and repositioned, it is best to remove all mortar and start over.

After the block is in position, the trowel needs to scrape off all excess mortar.

After several block have been laid, the mortar needs to be tooled. Tooling does several things. It makes the mortar joint look good; but more importantly, it densifies the mortar and reduces the likelihood that water will penetrate at the mortar joint.

Developing a Water-Resistant Wall

Two photos are shown of masonry work.

One wall was (1) laid with a good mortar, (2) using graded sand, (3) laid for even mortar joints, and (4) tooled at the proper time. This wall will resist the penetration of water when hurricane winds and rain blow against it.

Photo 15. Masonry wall. The mortar joints are all the same size, and each is tooled to consolidate the mortar. As a result, there are no hairline cracks between the brick and the mortar. This leads to a water-resistant wall.

The other wall was (1) laid with poor mortar, (2) contained sand that was not graded, (3) had uneven mortar joints, and (4) had joints that were not tooled at the proper time. Additionally, the block were not laid to produce the strongest column. This results in a weak wall. Since the grout in the column was not consolidated properly, there is a very weak joint as well.

One could say that the mason was not interested in quality work, but even the best mason, with the materials used, would have a difficult time producing a quality wall.

This wall will let water enter the wall and pass through the wall.

Photo 16.   Masonry wall.   With poor materials, it is difficult to produce a quality masonry wall. To improve masonry, not only is training needed, but the quality of the raw materials needs to be upgraded.

After the mortar has reached its initial set, the wall should be moistened to keep the mortar from drying out before the mortar hydrates.

Keep water out of the cavity after the wall has been laid. If water enters the cavity before the mortar has set, it can weaken the mortar or cause it to erode. If it enters the cavity after the wall has been placed into service, it may lead to water entering the structure. This can be done in several ways. The cells of the

Masonry

masonry units can be filled with grout or a beam can be poured over the masonry wall.

As with insulated concrete forms, both horizontal and vertical rebar can be placed in a concrete block wall. After the rebar is placed, grout (thinned mortar) can be poured into the cavities. This rebar/grout combination helps the wall resist tremors from earthquakes and provides a wonderful point to tie the roof to the wall. It must be well rodded to get it to the bottoms of the cells.

Some years ago, I ran a study concerning water passing through brick walls. I found that the water seldom passed through the brick, even though some of the brick tested would absorb large amounts of water. The water passed through the cracks between the mortar and the brick. By improving the bond strength of the mortar, the walls would become more water-resistant.

With concrete block, the situation is different. Water will pass through most concrete block. When concrete block are used for building a home, the block should be plastered both inside and outside.

Chapter 11

Stucco

Stucco Formula
A good mortar will work as a plaster or a stucco, but a little more sand will improve it.

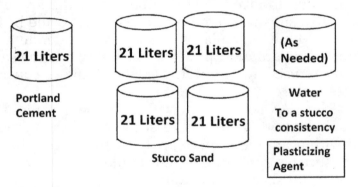

Stucco Formula

21 Liters — Portland Cement

21 Liters | 21 Liters

21 Liters | 21 Liters — Stucco Sand

(As Needed) Water — To a stucco consistency

Plasticizing Agent

Mix cement, sand, plasticizing agent, & water to a mortar consistency.
Sand - max. particle size 0.6 cm   0.4 cm is better
Round sand is better than crushed sand.
Discard batches not used within 45 minutes.
Re-temper not more than one time.
Moist cure for 3 days.

Figure 18.   Stucco formula.

Adjust the water content of the stucco so when holding the stucco on a trowel that is held at a 45-degree angle, it will stay in place. If tipped to a 60-degree or 75-degree angle, the stucco will slowly slip off. If the trowel is loaded and then flipped so the stucco is below the trowel, it will be held in place by the suction.

Plastering on Concrete Block

Some people like to stucco concrete block walls. Reasons range from aesthetics to serving as a moisture barrier. If the block are weak because they contain an excess of clay or because they were not vibrated during the manufacturing process, the wall may need to be plastered to keep it from failing.

If a decision is made to stucco a block wall, the mix design should be sand, Portland cement, and something to give the stucco body and the ability to stick to a substrate and hold onto it. That could be hydrated lime or plaster fat. See Addendum C concerning ground-up cactus. Basically, it is the same formula used for mortar used in laying a concrete block wall.

The day before the wall is to be plastered, it should be wetted down at least twice and then repeated within an hour before applying the stucco. Before the stucco is applied, the wall should be dried so there is no water sheen on the wall.

If the block are in good condition, there is no need to use lath; in fact, a better job will be accomplished if the stucco is applied directly to the concrete block. In the US and some other countries, a water-repellent or sealer is applied to the block during the manufacturing process. If this happens, the plaster will not bond well to the block. In that situation, block should be ordered which do not have the sealer, or lath should be applied to the wall to receive the stucco.

If the block are weak, then even though a stucco coat or surface bonding coat is applied and the wall looks good, it may not be structurally sound enough to handle a hurricane or an earth tremor. Before using low-quality block, ask, "Would I want to be in this building with my family if we had a hurricane or an earthquake?"

See the section on Masonry Units in Chapter 10 for a simple test for determining the quality of concrete block.

The stucco can be applied with a trowel or with a sprayer. If applied with a trowel, a throwing action will result in a better bond to the substrate than if it is pasted on with a trowel. A sprayer will throw the stucco at the wall so it will end up tight against the wall.

The first coat should be less than a centimeter thick. As it cures, the mortar lines between the block will probably show through the stucco. This is known as telegraphing. Wait 48 hours before applying the second coat. If the stucco is applied without waiting, the mortar lines will continue to telegraph through the stucco.

A second reason for waiting 48 hours before adding the second coat is that if the first coat of stucco cracks after the second coat is applied, the second coat will crack as well.

The second coat is about as thick as the first coat. It can usually go on smoother and flatter than the first coat since the suction of the wall has been evened out by the first coat of stucco.

On a block wall, the total thickness of the stucco should be about 1.25 to 1.5 cm. The third coat can be the finish/texture coat and is usually much thinner than either of the other coats.

Plastering on Lath

Lath is used in many different situations. Here are a few of them:
• Over a substrate with wood framing behind it,

- Over masonry where a good bond cannot be developed to the masonry,
- Over open framing where the lath is hung directly on wood or metal studs.

If there is not a firm substrate to apply the stucco to, often lath is applied. In many cases a felt paper or building paper is applied before the lath so water cannot penetrate through the stucco and enter the building. Since water usually gets behind the stucco at windows and other openings and at cracks in the stucco, a system needs to be installed so the water can drain out.

On the downwind side of a building, there is a suction. *See Figure 11 in Chapter 7.* The suction gets stronger as the wind blows harder. If the lath is not attached firmly, it can be peeled off the wall. How strong is "strong enough?" ASTM C1063 (Standard Specification for Installing of Lathing and Furring to Receive Interior and Exterior Portland Cement-Based Plaster) speaks of every 178 mm along each framing member when being attached to wooden framing members (studs). Those studs are usually spaced every 41 cm. That is one fastener every 730 cm². That is 12.7 fasteners per square meter. In areas of anticipated high winds, more fasteners are often required.

When using lath over concrete block, the lath reduces the pressure with which the stucco is pressed against the block; and as a result, one should not depend on any bond that might develop, even if there is no sealant on the block.

Plastering on an Airform

Plastering on an Airform is different. More information can be found in the book *Homes for Jubilee* (English) or *Kay pou Jubilee* (Haitian Creole).

After the Airform is in place and inflated, it needs time to stabilize. This is usually done by maintaining the air pressure overnight. It needs to be coated with a bonding agent. This helps the concrete or stucco to bond to the slick plastic of the Airform.

Starting at the foundation, a coating of stucco needs to be applied that is about 1.9 cm thick. Thickness is hard to judge, so build a thickness gauge by driving a finishing nail into a piece of wood that is nominally 10 cm x 5 cm x 2 cm. The nail should protrude 1.9 cm from the wood. This makes a handy gauge for measuring the thickness of the stucco coating.

Spraying the stucco on is better than troweling it on. There are some people who can apply the stucco by throwing it on with a trowel or shovel. The only troweling of the surface is to knock down high spots. If the surface is rough, better bond will be established when the second coat is applied.

Keep working up the side of the Airform. Do not put weight on the Airform, and do not let the pressure drop. Both will cause cracks in the stucco. After the stucco has gotten stiff enough so it can be dampened without washing any of it off, start dampening the stucco. The concept is not to add water to the stucco, but to keep the surface moist so moisture does not evaporate from the stucco.

After the first coat has been applied and cured overnight (keeping it moist), the dome should be wrapped with basalt rope or other approved material. Then a second coating of 1.9 cm of stucco should be added.

Again, it should be kept moist.

The following morning a third coat of approximately 1.9 cm should be added. This coat can be troweled so it is smooth or finished in whatever manner one desires. Stucco should be moist-cured for at least 48 hours.

Curing

Stucco needs water to cure. If it does not have enough water, the stucco will not form chemical bonds to the substrate.

During the curing process, keep the surface of the stucco moist for at least 2 days. The concept is to keep the moisture from evaporating from the stucco so the internal moisture can react with the cement particles in the mix.

If the stucco turns from the rich gray color when it is first applied to a light gray color, that means that it has been allowed to dry out. Don't let that happen.

A light mist on a regular basis is much better than a hard blast of a hose periodically.

Limitation of Stucco

Stucco should not be considered a structural layer. If the building is not designed to survive a disaster, a coat or two of stucco will not help. The exception to this is when building a dome or when using the stucco as part of a ferro-cement structure.

> *Author's Note: With Hurricane Matthew, several buildings which were supposedly disaster-resistant failed. They consisted of a metal frame, then wood was attached to the metal frame, and then lath was attached to the wood. Both sides of the lath were plastered. Since some of the stucco contained excess clay, the bond of the stucco to the lath failed. Since undersized nails were used to hold the lath to the wood, in many areas the lath tore away from the wood. Since the same people poured the concrete floor for these houses as applied the stucco, two of the concrete slabs washed away.*

Quality Concrete from Crap

Chapter 12

Surface Bonding Cement

Surface bonding cement (SBC) and the resulting mortar from mixing SBC with appropriate sand was developed to bond concrete block together with a stucco-like substance which has an adequate bond strength and is adequately fiber-reinforced. The resulting wall has block which are bonded together to form a monolithic and stable wall. The need for a craft mason is avoided. The stacking just needs to be straight-linear and vertically plumb.

Background

After an adequate foundation has been poured and cured, concrete block (officially known as Concrete Masonry Units-CMUs, but they will be referred to as block in this document) are dry-stacked without space between the block on any row. Block are stacked with no joint space either horizontally or vertically. Some of the block may not be perfectly shaped and may tend to rock when placed. To correct the potential movement, wedges or other means are used to stabilize the block. Since the walls may not remain vertical as they are dry-stacked, more wedges are inserted.

Dry-stacking and surface bonding of walls is a fairly rapid way of building strong walls. As with any building technique, there are construction techniques which result in stable walls; but if attention is not paid to the surface bonding cement mortar used and the way the walls are constructed, unstable walls can result.

The walls must be braced in place during construction if any potential of them falling is possible. Dry stack block techniques and tools are discussed on the website – www.drystacked.com.

Essentially, the technique is to dry-stack concrete block without mortar and then apply a thin layer of surface bonding cement mortar to each side of the block wall. After the mortar has cured for 24 hours, selected cores in the block wall are filled with rebar-reinforced concrete. If a one- or two-story block house is being built, the corner cells are filled, the cells on each side of each window and door opening are filled, and cores are filled every 1.2 meters.

The system works better if the block are manufactured in a modern concrete block factory (so all block are of standard size and shape). Rather than cutting block, it is better to purchase manufactured half-block.

The system can be modified to use with confined masonry construction.

The major advantage of this system over mortared block is that the block can be dry-stacked much faster than mortared block can be laid, since each block does not have to be set precisely to get a straight wall. After several block have been stacked, they can be aligned with a string line and a rubber hammer.

A disadvantage occurs if the block are not uniform; the wall may become unstable before it is stabilized with surface bonding cement and by filling of cores. A second disadvantage occurs if care is not taken to install the surface bonding cement to produce the greatest bond strength.

Surface Bonding Cement

Foundation

The same steps should be taken in developing the foundation as when developing the foundation for any concrete block wall. Where disaster conditions may occur, reinforcement should come out of the slab/footings and extend not less than 60 cm above the surface of the slab. Ideally, the location of the foundation rebar is the same as the cells which are selected to be reinforced and grouted (see Reinforcing the Wall below). When these cells are filled with concrete, this will prevent the structure from shifting on the foundation in the event of an earthquake.

Block

Any concrete bock can be used, but it is better if they have been made by a modern concrete block plant, so they are all the same shape and size. They also need to be structurally sound. If they are not all uniform, it becomes difficult to dry-stack a wall which will be stable. If they are not uniform, periodic shims can be placed between block to ensure that they do not rock and that the wall remains vertical and flat.

Leveling Courses

Leveling courses, when needed, are to provide a smooth surface, level within 0.5 cm per 5.0 meters, for dry-stacking block. Leveling courses should be located whenever a vertical difference greater than 2.2 cm in 5.0 meters occurs within one course. Leveling courses are usually located on the first course above the foundation (because footings may not be placed in a level condition) and at each floor level.

When mortar is used as a leveling course, block are set in a full bed of mortar, laid to a line with the top surface level and butted together with no mortar in the head joints. Bed joints are struck flush. If the cores are to be grouted, no mortar is placed in the space to receive grout.

Allow the leveling mortar to set sufficiently so no movement breaks the bond when dry-stacking units in subsequent courses.

Dry-Stacking Block

Courses of block between the leveling courses shall be placed without mortar on the bed or head joints. Place units in a running bond (overlapping) pattern. Remove burrs and butt blocks tightly.

Use shims, mortar, or surface bonding mortar to plumb and level individual units when necessary.

Check the wall every fourth course to be certain it is plumb and level. If any course is out of level by more than 2.2 cm in 5.0 meters, another leveling course is needed.

Cut block to fit openings. Minimum length of cut pieces used in the wall must be 3.2 cm. Anchors, reinforcement, flashing, lintels, and other items to be built are installed as the stacking progresses. Cut or notch block as required.

Wiring a Home

Utilities such as electrical lines and plumbing located in the cores of the units are best placed prior to the application of surface bonding mortar while the block are visible.

By cutting the webbing in the block, conduit can be inserted into the walls and electrical boxes can be mounted. The other common option is to mount the conduit and electrical boxes on the surfaces of the walls.

String Lines

To lay the walls straight and vertical, it is appropriate to start with carefully positioned story poles. Between the story poles, string lines are placed to ensure that the block are laid where they need to be laid. As with mortared block, the block are not laid touching the string lines, but a specified distance from the string lines.

The first tier of block should be laid on a mortar bed to ensure that the tops of them are all level and that they do not rock. Space should not be left between the block as if a head joint were to be applied.

After several tiers of block have been laid, the block need to be shimmed to ensure that the top block are level along the length of the wall.

Surface Bonding Cement

In some countries surface bonding cement is available at building supply centers; but in most third world countries, at the present time one must make his own.

Following is a recipe for a 12-kg batch of Surface Bonding Cement Mix.

- Portland cement   8.26 kg
- Hydrated lime (Type S)   3.6 kg
- Glass fiber (1.27 cm)   140 grams

Normally, the bonding mix is applied to a minimum thickness of 0.32 cm on both faces of the block wall.

Surface Bonding Cement Ingredients

Portland cement is available worldwide.

While Type S dolomitic hydrated lime is preferred in the mix, it is not as available in third world countries as Portland cement.

If hydrated lime is not available, finely ground limestone may be substituted. Workability is not as good, but workability can be improved if 12 grams of Methocel (if available) is added.

Methocel is a commonly available admixture which is readily available in first world countries. If large amounts of surface bonding mix are to be made and hydrated lime is not available, Methocel should be imported.

Glass fiber serves as reinforcement. There are now a number of other fibers on the market which will serve as well, including poly fibers, nylon fibers, and basalt fibers. They are used to provide reinforcement, so if stress is applied to the wall, the thin coating of the bonding mix will not fracture. Fibers are becoming more and more available in third world countries. The ideal length is about 1.27 cm. This length provides the needed strength, and when mixed in a mechanical mixer, will not tangle as much as longer fibers. Finer fibers provide more reinforcement than coarser fibers.

Surface Bonding Cement Mortar

Surface Bonding Cement Mortar can be made by adding 12 kg of fine sand to the above mixture. The sanded mix is a much better gap filler, so it is especially helpful if the surface of the block have pits and if there are gaps between the block. As with the surface bonding cement, it should be applied 3.2 mm thick on the surface of the wall.

The fine sand used should pass through a 30-mesh sieve. That is 30 wires per inch, or about 12 wires per centimeter. That means the particles are smaller than 0.6 millimeter. Hard silica sand is preferred. The sand should be washed so it does not contain any clay. Aluminum window screen wire is usually about 27 mesh.

Mixing the Bonding Mix

When making the mix or the mortar, all dry components should be thoroughly blended before water is added.

To obtain the maximum bonding, and this is what the product is designed for, the material should be used within 30 minutes of mixing. As a result, it is better not to mix more than 11.36 kg (25 pounds) at a time per applicator.

While greater compressive strength can be obtained by limiting the amount of water, to obtain maximum bonding more water is needed. It is best to mix to the most liquid consistency possible and have it still stay on the wall without signs of sagging. This will result in a consistency about like toothpaste.

Applying the Bonding Mix

Block should be free from dirt and any other items. The bonding mix should be bonding to the block, not to loose paint, dirt, or other items.

Spray the wall down several times before starting to apply the bonding. It should be damp, but not shiny, which would indicate that there is unabsorbed water on the surface. Wet the wall uniformly with water immediately before applying surface bonding mortar, to prevent excessive suction of water from the surface bonding mortar. If the wall dries prior to application, rewet it. Avoid saturating the units.

The bonding should be applied to both sides of the dry-stacked block with firm pressure. Better yet is to use a mortar sprayer to blow it onto the wall. This gets the bonding mix tight to the block better than troweling can. Trowel or spray-apply surface bonding mortar to both sides of the wall. Completely cover the wall surface with a minimum thickness of 1/8 in. (3.2 mm) of surface bonding mortar by trowel or spray.

Quality Concrete from Crap

When a second coat or color coat of surface bonding mortar is to be applied, the first coat shall have taken its initial set but not be completely hardened or dried out. If the first coat has completely hardened or dried out, consult the manufacturer's recommendations for application of a second coat, since a bonding agent may be required. Finish to the texture specified.

If application of surface bonding mortar is discontinued for more than 1 hour, the horizontal joint between the two applications, other than at wall tops, must occur at least 3.18 cm from the horizontal edge of any block.

If the block are flush with the outer edge of the slab, the bonding mix should carry down along the side of the slab.

For best results, start applying the bonding mix 0.6 to 0.9 meters from the top of the wall and then work upward. After that is finished, drop down and select another section of about 0.6 to 0.9 meters high to work. Continue to work until the bonding mix is in contact with the foundation upon which the block are laid. By working in this manner, the wall can be re-wetted without water running over the fresh bonding mix.

Essential Steps

Apply the bonding mix with firm trowel pressure, pushing the load upward until uniform coverage is attained.

Then using longer, lighter strokes, and holding the face of the trowel at a slight angle, smooth the surface.

Move to the area below and continue.

With the trowel moist, retrowel the area with long strokes while holding the trowel at a slight angle to smooth out any unevenness.

Surface Bonding Cement

Over-troweling brings fines to the surface and results in hairline cracks and crazing. Ideally, the surface should have a slightly fibrous texture. Such a surface is less likely to develop a hairline crack than a smooth, over-troweled surface.

Think of the bonding mix like duct tape. If it is applied on one side of a block wall, the wall has no resistance to pressure applied to the opposite side of the wall. When duct tape is applied to both sides, the wall becomes resistant to forces from either side.

Reinforcing the Wall

Once the block walls have been surface bonded, vertical cells in the wall should be selected to receive reinforcing. As a minimum, the corner cells should be selected, at least one cell every 1.2 meters, and alongside each window and door opening. It is better if three cells are selected to fill at each corner.

Each selected cell should receive one piece of #5 rebar or two pieces of #3 rebar. The rebar should extend above the height of the wall to tie to the second floor if a second floor will be added or to tie to the concrete roof if a concrete roof is to be added. If a truss roof is added, the rebar need not be as long, and bolts can be inserted into the top of the concrete mix to tie the top plates to the walls. Then a concrete mix should be poured into each selected cell. Maximum aggregate size should be 2.0 cm. The rebar can be used to rod the concrete as it is being poured into the cells.

Fill the corner junction between the wall and footing, carrying the bonding mix onto the top of the footing on both sides of the wall. If the wall is built on a concrete slab floor on grade, carry the surface bonding down over the outside edge of the slab to help seal the joint between wall and floor.

Quality Concrete from Crap

Cracking can occur if the walls are not kept moist. Wet the finished bonding mix with a fine spray of water two or three times a day for a couple of days to aid the curing process. Then wet the walls down once per day for 3 or 4 weeks.

Finishing the Building

The exterior and interior sides of the bonded walls need to be waterproofed. A high-quality, non-porous latex works well. If Methocel is used in the mix, it provides waterproofing.

Roof construction can begin a week after the bonding is completed.

Chapter 13
Roofs

Non-Concrete Roofs

After numerous concrete roofs collapsed after the 2010 Haitian earthquake, there are many people who will not consider living under a concrete roof. The problem is not concrete roofs, but the poor quality of many concrete roofs

Whatever type of roof is to be used, there needs to be a concrete bond beam all the way around the building so the roof can be solidly anchored to the walls. There needs to be a method so that each roof truss or rafter is firmly anchored to the bond beam.

Since non-concrete roofs do not offer the bracing between opposite walls that a concrete roof will offer, steps need to be taken to install bracing. It can be in the form of concrete bond beams. It can be in the form of concrete bond beams on top of dividing walls within the building.

While wooden trusses can be built which will reduce the pressure applied to the walls below them, they are not as good as concrete beams, in my opinion. Sometimes there is no choice, for example in a church where a large open space is desired below the roof. In that case, buttresses can be built on the outside of the building to provide bracing. Gothic cathedrals in Europe used buttresses, and later they used what became known as flying buttresses. These were ornamental structures which also provided support to the walls.

Metal trusses can be built which are much stronger than wood trusses.

> *Author's Note: In October, 2016, following Hurricane Matthew in Haiti, I examined a church which lost its roof, and one wall was seriously cracked. The gable roof did not have an adequate bottom chord, and there was a downdraft. The roof was pushed down, and that made the ends spread. The wall, a beam, and a column were cracked.*

Photo 17. The wall of a church that lost its roof. Contributing to the problem was that the column and beam were not adequately reinforced.

> *Author's Note: In the 1950s at a cemetery outside of Roma, Texas, was an entrance structure consisting of two brick columns and a very low arch between them. The columns were one brick wide. Each time we would pass it, my father would ask me to explain why the structure did not collapse, since there was more lateral pressure from the arch than a column of brick could resist. The first time I was driving past the cemetery without my father, I stopped. There was a thin stainless-steel wire just below the brick arch tying the two columns together. It could not be seen from the highway. Be innovative, but address all stresses which are likely to be placed on the roofs and walls of any building you help construct.*

A hip roof with short eaves is probably the most-disaster-resistant and easiest-to-build roof to place on a disaster-resistant home. With these roofs, if the building is square, the roof is peaked, and the same slope is used on each side of the roof. Each side of the roof is identical. If the building is rectangular, there is a ridge line, and the slope for each side and end of the roof is identical. The reason the overhang at the eaves should be limited is so wind cannot get under the eaves and tear the roof off.

The next best roof is a Dutch hip roof. It looks like a hip roof, but there is a vent at each end of the ridge line. This can be developed by extending the ridge line or can be developed by having a flatter slope on the two ends. The advantage of the Dutch hip roof is it allows ventilation in the attic space which keeps the attic cooler and thus keeps the building cooler.

Quality Concrete from Crap

Gable roofs normally provide large areas at the ends where hurricane force winds can tear the roof off. If a gable roof is used, the overhangs at the eaves and the gable ends should be limited so wind must work harder to tear the roof off.

The worst non-concrete roof is a shed roof. Even with a short overhang, the force lifting the roof on the upper edge is tremendous if the wind is coming from that direction. Remember, depending on where the eye of the hurricane is located, the wind can come from any direction.

Flat Concrete Roofs

Flat concrete roofs are easy to build, but are easier to build wrong. When supported by the walls of the structure and not providing a substantial overhang, most of the reinforcement needs to be in the lower portion of the roof slab.

A 10-cm-thick flat roof on a 6-meter-by-6-meter house will weigh about 9,100 kilograms. This requires substantial support. If it is poured in one piece, it requires a substantial frame to support it while the concrete cures. Such a roof will fail. It needs to be 20 cm in thickness.

A flat concrete roof should never be flat. It should have a slope to it. Different people have different recommendations, but most of them are based on the concept that once the concrete has shrunk, there will be areas where water will pool. The slope should be enough so these pools will drain. 2 cm per meter is normally adequate. If the concrete is porous, the shrinkage will be greater and the likelihood of water seeping through the roof slab is greater.

Conventional Flat Concrete Roof Construction

Reinforcement of Simple Roof Slab
View 1

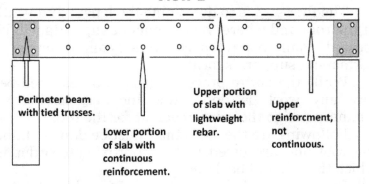

Figure 19. Reinforcing a simple roof slab. The perimeter beam surrounds the slab, and all rebar are tied to the perimeter beam. Roof should slope about 2 cm per meter.

Reinforcement of Roof
View 2

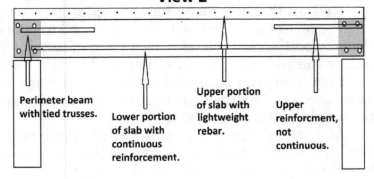

Figure 20. Reinforcing a simple roof slab, as seen from a right angle to Figure 19.

Quality Concrete from Crap

A concrete roof can be constructed so it stops at the walls which are supporting it, or it can be constructed so that there is an overhang which protects the exterior walls. Building with an overhang produces stresses on the roof that are not present when a roof is built without overhangs.

The first step is to determine the length and width of the roof and where the permanent supports will be located. Plan to have a concrete beam wherever a permanent support is located.

While the entire process needs to be planned before any work is done, this includes having an engineer design the reinforcement for the roof.

Following are the steps that must be done to build the roof and are listed in the approximate order in which they should be done.

Install the reinforcing trusses for the concrete beams which are resting on the support walls.

Install the decking which will support the roof while it is being poured. Normally this is plywood. Boards need to be attached on the underside to hold the pieces of plywood together. The plywood needs to be supported. Vertical 2x4 lumber can be used. A block needs to be placed under each piece to spread the load on the slab. Between the top of each piece and the plywood, a block needs to be placed to spread the load on the plywood. Having a piece of 2x4 poke a hole in the plywood leads to serious problems.

Tree limbs can be used for the vertical supports. Each needs to be tested to ensure that it will support the weight that is placed on it.

After the decking is in place inside the building (and extending outside the building if the roof is going to extend past the walls), the decking needs to be coated with a bond breaker. Polyethylene plastic is one option. Coating the plywood with used motor oil is another option.

Now the reinforcement across the short span of the roof needs to be installed. Reinforcement needs to go only where the concrete will be under tension. That means high over supporting walls and low between supporting walls. The amount of reinforcing materials could be limited if one could be sure that it stayed in the correct location as the concrete is poured. On small jobs it often does not.

An option is to build trusses from the reinforcing material and attach those trusses to the reinforcing materials in the beam trusses. Ensure that the lower chord of the reinforcing truss is at least 2.5 cm above the bottom of the roof slab and that the upper chord of the reinforcing truss is at least 2.5 cm below the top of the roof slab.

Another option is to place continuous reinforcement where the bottom chord of the trusses would go and then add shorter pieces which are tied to the beam trusses and extending out from them to ensure that there is reinforcement wherever the concrete slab will be under tension. An engineer designing this is well worth anything he must be paid.

Above the reinforcing trusses, a lighter-weight reinforcing should be placed at right angles. These can be placed 2.5 cm above the upper chords of the roof trusses, or if the upper chords are not continuous, they may be supported so they remain in place. The trusses are there to carry the weight. The right-angle reinforcing is there to reduce the cracking of the roof slab due to temperature variation.

At this point, everything should be checked and rechecked. Once the concrete pour is started, there is not time to make changes. Ensure that there are enough materials to produce all the concrete needed.

Ensure that there are enough people to mix and transport the concrete to the roof. Ensure that there are enough people to rod or jitterbug the concrete to ensure that it is well-consolidated.

It is a common practice to place concrete block in between the reinforcing to take up space, so not as much concrete is needed for the pour. This procedure usually leads to problems a few years down the road. It is better to use the system listed later in this chapter to build lightweight block from lath and a little stucco.

Once the concrete pour has started, do not stop until it is finished. Having an extra mixer is important. Having extra of everything is important.

After the concrete has been poured and the water rises, it should be finished. Keep it moist for at least three days. Keep the support posts and decking in place for at least seven days.

Lightweight Concrete Roof Construction

The Blondet Manual shows a lightweight concrete roof using lightweight hollow clay tile. Such tile is not made in Haiti and would be too expensive to import. Several paragraphs are included which give concepts for ways of replacing the lightweight tile with a manufactured block.

> *Author's note: The English word **tile** refers to a "slab or block" made from clay and fired in a kiln. The English word **block** is a generic term which includes ceramic tile.*

There are several ways to make lightweight block. They need to be strong enough to hold their shape when boards nominally 2.5 meters long and about 15 cm wide are laid on them and one or two people walk on the boards. Then they need to be strong enough to support a 5-cm layer of concrete over the top. Here are several options:

Styrofoam could be used to replace the lightweight block, *BUT* it could float in the fresh concrete. Lath would need to be added under it so the ceiling could be plastered.

Cardboard could be used and sprayed with Thompson's Water Seal so it would not soften when the concrete is poured next to it, *BUT* it would need to be stiffened to hold up with the weight of the crew while the concrete is being placed.

Papercrete block can be made from waste paper and cardboard that are slurried and poured into molds and allowed to dry, BUT it would be difficult to keep it dry.

The builder could make box forms from lath and use wood to reinforce them, or lightly stucco them to reinforce them.

A final method would be to construct the block from stucco lath and a little stucco. The next section, Lightweight Block Production, discusses producing the stucco lath and stucco block.

The lightweight roof consists of alternating beams and lightweight block with a slab poured over the top of it.

The beams are 10 cm wide (4 inches) and 15 cm high (6 inches).

The lightweight block are 30 cm (12 inches) wide and 15 cm high (6 inches).

Over the top of the beams and lightweight block is a 5 cm (2 inches) continuous slab.

Rebar are placed as needed, based on the spans between walls.

Solid decking can be installed, or an acceptable way to proceed is to use strips of decking. If that is

Quality Concrete from Crap

done, lumber that is at least 15 cm wide can be laid on supports on a 40-cm center-to-center basis. This allows each block to rest on at least 2.5 cm (1 inch) of the boards. Care must be taken to not inadvertently move the lightweight block before all the concrete is in place. In the following depiction, rebar are not shown.

Lightweight Roof Concept

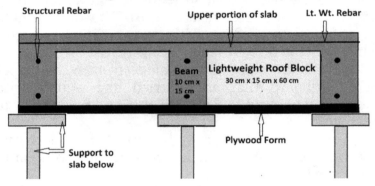

Figure 21. Lightweight roof concept. Once the roof pour is started, it must be completed in one operation.

A continuous pour is used to pour the perimeter beams, the roof beams, and the 5-cm slab, so there are no cold joints. During the pouring of the beams, walking boards are placed on the lightweight block to support the crew pouring the concrete.

Lightweight Block Production

Following is a description of a lightweight block I have made. They could be readily made by home businesses. This description is for a block which is 15 cm thick. The dimensions can be adjusted for the desired thickness of the roof.

Lath is 69.6 cm (27 inches) wide. The blocks need to be 15 cm high (for a 20-cm thick roof) and 30 cm wide, so pieces of lath 108 cm long (plus lap) could be cut and formed into block that are 69.6 cm long.

1.25 cm of stucco on the sides would support the walking boards on top of the block which the men stand on as they pour the concrete. On each end of each block, 3.8 cm of lath would be folded at right angles and tied. This would keep the block from collapsing to the side, especially if a little stucco were dabbed at each corner.

Lightweight Roof Block

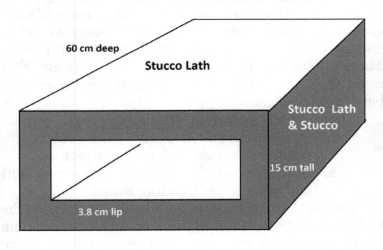

Figure 22. Lightweight roof block made from stucco lath and stucco. Height and width can be varied, depending on needs.

That would make a 60 cm (24 inches) block and would provide a rim on each end 3.8 cm wide to stucco. That should be strong enough to hold the men pouring concrete, especially if the lapped stucco were wire-tied at each of the 8 corner laps.

Such a block would weigh about 6 kg, and it would be hollow inside the lath from which the block was formed. This cavity would provide space for running electrical and other lines through the block. With a microloan, a business could be set up so this type of block could be produced and furnished to the builders.

Roof Produced from Concrete Strips

Rather than trying to pour a complete roof with one pour, one can pour strips of reinforced concrete and hoist them up to form the flat roof. If the strips were 20 cm thick, 30 cm wide, and 3 meters long, each one would weigh about 440 kilograms. With enough people, these can be lifted and slid into place. Then a 5-cm slab is poured over them to ensure that water does not enter the building through the gaps between the strips.

There are a couple of problems.
1. Due to movement of the strips, cracks develop in the 5-cm slab and let water in.
2. During an earthquake, the strips can shift and fall into the building below.

Unless there is a secure way to tie the strips to the walls, unless the walls are strong enough to handle the lateral movement, or unless one can ensure that there will not be any earthquakes or tornadoes, this roofing technique should not be used.

Ferro-cement Roof Channels

This is a way to build lighter-weight roofs, but without all the support structure. Ferro-cement, mentioned earlier, can be used to form roof channels, which are stronger than a flat piece of concrete.

Essentially, a semi-circle mold with a lip attached is used which is as long as the roof is wide. In cross section the channels look like an "S" with the upper curve very small and the lower curve very large. The mold is about 80 centimeters wide and about 40 centimeters high. The mold is covered with plastic as a bond breaker and then coated with a centimeter of stucco. Wire mesh is laid over it, and the stucco is built up until it is about 2.5 cm thick. A channel that is 6 meters long and covers a span of 80 cm will weigh about 450 kilograms. A roof for a 6-meter-square house would weigh about 3.5 tons, about a third of the weight of the 10-cm-thick flat roof, and would be much stronger.

Ferro-Cement Roof Channel Concept

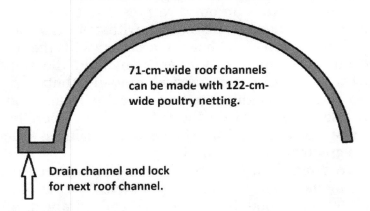

Figure 23. Ferro-cement roof channel concept.

There are companies which make molds and vibrators to make the ferro-cement roof channels.

Long-term testing of ferro-cement roof channels using SpiderLath as reinforcing needs to be done. SpiderLath does not deteriorate in less-than-ideal concrete like metal lath does. This is something which should be investigated.

As with the concrete strips, with enough people these can be lifted and slid into place. A 5-cm slab does not need to be poured over them to ensure that water does not enter the building through the gaps.

During an earthquake, the strips can shift and fall into the building below. Unless there is a secure way to tie the channels to the walls, unless the walls are strong enough to handle the lateral movement, or unless one can ensure that there will not be any earthquakes or tornadoes, this roofing technique should not be used.

Arched Roofs

An arched or vaulted concrete roof is harder to build, but it can be built stronger than a flat or gabled roof. Since there is an outward pressure, the walls of the structure need to be built to withstand that outward pressure. The higher the arch or vault, the less the outward pressure. Rather than build the walls thicker to handle the outward pressure, buttresses were added in Gothic buildings. Steel wire and rods can also be used to tie the two walls together. Remember, steel rusts and corrodes, so stainless steel has a greater chance of surviving.

To build an arched roof, a form as wide as the span of the roof and about 1.5 meters long, is often built from plywood. It is positioned and shimmed. After the concrete arch is built over the form, the shims are removed and the arch is moved down about 1.5 meters, re-shimmed, and the next section of the roof is built.

A problem with using this method is that every 1.5 meters there is a cold joint in the concrete. Adequate reinforcing needs to be added, so adjacent segments of the roof can be tied together.

A balloon form can be constructed and inflated to provide the support for building the arched roof. The problem is ensuring that the area below the balloon form is airtight.

Chapter 14

Building Disaster-Resistant Homes

This chapter is a survey of several different types of structures which can be built to be disaster-resistant.

Be Prepared for Disasters

Hurricanes and typhoons are a fact of life in many coastal areas.

Wherever fault zones occur, earthquakes may occur.

Tornadoes frequent the Great Plains of the US and other areas of the world.

Forest fires and prairie fires can also bring about disasters.

Wherever rain falls, floods may occur.

In areas which experience periodic disasters, it is appropriate to build to withstand those disasters.

Think about the whole structure and how it will react to a disaster. In California, many homes with fire-resistant walls and roofs have burned because embers entered the attic through soffit vents. The wood in the attic was very dry, so it was easy to start a fire. If soffit vents with finer screens had been used, the houses would not have caught fire.

With floods, the easiest way to avoid them is to build above the highest known flood line. In the US, most water courses have been mapped to determine the maximum expected flood in 25 years and the maximum expected flood in 100 years. Periodically, events occur which demonstrate that those are just best estimates from known data. Hurricane Matthew in Haiti and Hurricane Harvey in the US demonstrated this. Both stalled out and brought more rain, and thus more flooding, than was anticipated.

When homes are destroyed by one of these events anyplace in the world, there is an effort to rebuild. All too often, the rebuilding is to the same standards that were used to build the structures that were destroyed.

In 2008, Hurricane Ike destroyed expensive houses on Galveston Island and on the Bolivar Peninsula in Texas. Even though there was a SCIP home that survived without damage while other houses were destroyed, people rebuilt so the next major hurricane would wipe them out again. Hurricane Harvey, in 2017, was the next major hurricane.

Building hurricane-resistant and earthquake-resistant homes is not much more expensive than building homes that will be blown away within the next 10 years. The secrets of building disaster-resistant homes are easy to learn.

The Disaster Experts

Ken Luttrell and Joe Warnes probably know more about disaster-proof homes than just about anyone else. I know them from World of Concrete (an international concrete trade show with seminars, workshops, and panel discussions), where all three of us have been speakers. A few years ago, I spent an evening with them, along with three others who are as knowledgeable about things concrete as I am. Also present were:

Dave Stevenson from Advanced Structural Panel Industries, LLC, who is an expert on Structural Concrete Insulated Homes,

Jesse Lilligren, who has built in disaster areas all over the world, and

Nolan Scheid, owner of MortarSprayer.com.

While we all are considered experts in our little niches, we found that Ken and Joe knew much more than we did. They explained that one should not develop specifications for building earthquake-proof structures and another specification for building hurricane-proof structures, and yet another specification for building firestorm-proof structures. If a home were built to resist a Class V tornado, it would resist all other disasters except for a meteor strike, a direct atomic blast, or the earth opening and swallowing the house. This results in having one design standard, which simplifies the design and construction process.

A Tornado-, Hurricane-, and Earth- quake-Proof Home

Ken and Joe's method of achieving a tornado-proof house is simple: build a six-sided reinforced concrete box that is tied together. Walls need to be a minimum of 10 cm (4 inches) thick and have considerable internal bracing (interior concrete walls).

To do this, one should establish a ground anchor. This will usually be a slab with footings under it. It should be built to withstand the loading of the walls and roof, as well as the added loading that can come from the wind. In most cases, one does not need to worry about the wind turning the structure over, but if it is extra high and the slab is not tied to the earth, that could be a problem.

If there is no concrete slab, but a dirt or gravel floor, one can still build a tornado-proof home. Enough holes need to be dug deep, filled with concrete and steel to provide the ground anchor. Often it is easier to install an adequate slab.

Quality Concrete from Crap

The concrete walls need to be tied to the ground anchor. This is usually done by tying rebar that extends up into the walls to the rebar mat of the slab.

The concrete roof needs to be tied to the concrete walls. This results in a monolithic six-sided box.

Holes are left in the box for windows and doors. Even if all the windows and doors were to be destroyed, the structure would still stand and resist the disasters that nature can throw at it.

Little things can make it a lot better. Provide drainage. Water should flow away from the house. If it collects around the house and the water gets deep, water will enter the house. Moderate the eaves. Long eaves give the wind something to hold onto and try to break the connection between the roof and the walls. Adding storm shutters can protect any window glass that may be present.

In talking to several structural engineers, they suggested that a better term than "disaster-proof structure" would be a "disaster-resistant structure." Typhoon Omar in 1992 wiped out 75 to 90 percent of the homes on Guam. That is when the US government went in and started building homes which would handle storms with over 200 mph winds. Since tornadoes have winds well over 200 mph, disaster-proof is not an accurate term.

Many houses in Haiti and other parts of the world are built from concrete, but most are not fastened together to make them disaster-resistant. That would require a few changes. A contractor with a reputation for building disaster-resistant homes will have more potential jobs than one who builds cheaply.

Here are a few other methods of building a concrete home that is disaster-resistant.

Domes

The domes we are talking about are concrete structures which do not have any straight sides. As mentioned before, strength can be developed by having a curved surface. While there are several ways to build domes, the easiest way is to inflate a balloon (Airform), then add rebar and concrete. The rebar and concrete can be added to the inside and the Airform remains in place as waterproofing, or the rebar and concrete can be added to the outside and the Airform is removed and used again. Normally, larger domes are plastered from the inside, and smaller domes are plastered from the outside. Both systems have advantages and disadvantages.

The books *Homes for Jubilee* and *Kay pou Jubilee* contain information on building domes where the concrete is placed on the outside of the Airform, which is then removed and used again.

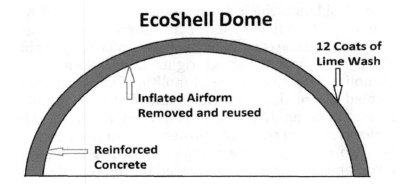

EcoShell Dome

12 Coats of Lime Wash

Inflated Airform Removed and reused

Reinforced Concrete

Figure 24. EcoShell dome.

With the larger domes, a urethane foam is often applied to the inside of the balloon before the rebar and stucco are placed. This results in the internal temperature of the dome approaching the temperature of the earth under the dome—never as hot as a conventional home in summer and never as cold as a conventional home in winter.

Monolithic Dome

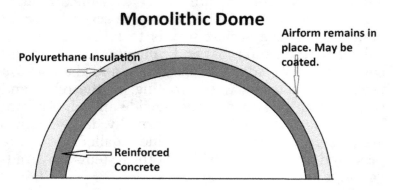

Figure 25. Monolithic dome.

With a concrete slab and with the dome tied to the slab, these structures weather all the disasters with ease. Problems which are periodically mentioned are that they do not look like conventional houses and that the space next to the walls may not be fully usable if they are not designed right. David South with Monolithic developed the Monolithic Dome, which is formed by erecting an Airform and first spraying urethane foam on the inside and then concrete. The Airform stays in place and protects the urethane foam. He also developed the EcoShell which uses an Airform, and the concrete is sprayed on the outside. The Airform can then be removed and used again for up to about 100 times.

Several years ago, a company developed a modification of the dome-building system that adds a front to the building that makes it look more like a conventional home, but it still has the disaster-resistant qualities of the concrete dome. I have not seen any advertisements concerning that company lately.

There is a learning curve to building a dome, but it is not a difficult process. A 15-year-old grandson decided he needed to move out, so on his parents' property he is building a six-meter diameter dome.

Domes, while they do not look like conventional homes, are the most economical of the disaster-resistant building techniques.

Structural Concrete Insulated Panels (SCIP)

Wire mesh is on both sides of the wall. The wire mesh needs to be imbedded in 2.5 to 3.0 cm of stucco. Wire trusses provide tensile strength between the sheets of wire mesh. The interior concrete for the wall can be replaced with foam insulation board.

A panel with 6 cm of stucco and 24 cm of foam insulation board behaves like a solid reinforced concrete wall that is 30 cm thick.

The panels can be used for walls, roofs, and upper floors on multi-story homes. The insulation helps keep the home cool.

Structural Concrete Insulated Panels
View from above

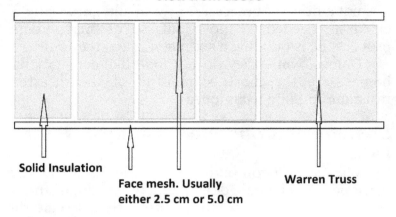

Solid Insulation

**Face mesh. Usually
either 2.5 cm or 5.0 cm**

Warren Truss

Figure 26   Structural Concrete Insulated Panel concept.

Warren Truss

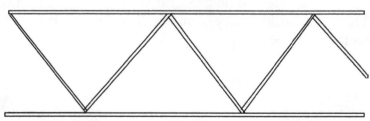

For SCIP, members are usually galvanized steel wire.

Figure 27.   Warren truss.   For SCIP they are usually made from galvanized steel wire.

> *Author's Note: About 20 years ago I inspected a subdivision built with the system in 1970. 30 houses were built. On 29 of them I could not find any cracking. The 30th house had been too close to the cliff edge, and during one of the California mud slides, it slid 25 meters down the cliff. When it came to rest, the doors would still open and close.*

SCIP structures are very robust. So far, no matter how shoddy the workmanship, I do not know of a single SCIP structure which has failed.

Homes can be built so they are disaster-resistant. I inspected a SCIP home following Hurricane Ike on the Texas Coast. Other homes were destroyed except one that was protected from the wind by the SCIP home. One home broke free from its foundation and crashed into the SCIP home. It broke into pieces. The SCIP home survived with minor damage.

While many standards call for these homes to be coated with a shotcrete (a concrete that is sprayed on, and has a compressive strength of at least 17.25 MPa), testing has shown that if the coating is at least 6.9 MPa, the homes are disaster-resistant. It is harder to make a plaster that has such a low compressive strength than to make one with the higher compressive strength.

Most of the plants around the world which produce SCIP cost more than five million dollars US to install. There are plants in the Dominican Republic, Central America, Mexico, Venezuela, Africa, and Asia.

Dave Stevenson, Advanced Structural Panel Industries, LLC, has designed and has completed the installation of a plant in Southern California that can be built for about 1.3 million dollars US.

It is an ideal technology for building disaster-resistant houses. The panels, usually 122 cm wide, are assembled in a factory. Picture a piece of rigid insulation with wire mesh on each side and with wires stuck through the foam insulation at different angles. The wire mesh on each side is not tight against the insulation but out from it about a centimeter or two.

The building process is straightforward. A slab is poured, with two rows of rebar sticking up every 0.5 meters. The panel is placed between the rebar, and the wire mesh is wire-tied to the rebar. Each panel is wire-tied to the next panel with a strip of wire mesh bridging the joint between the panels. The roof is tied to the walls with wire mesh bridging the joints. Panels are braced and then stuccoed. While some will apply stucco with a trowel, often a stucco sprayer is used. In areas where one does not want to invest the money to purchase a contractor-grade stucco pump and sprayer, MortarSprayer.com has a handheld sprayer that is operated by compressed air and filled by dipping it into a wheelbarrow of stucco.

Because of the way the panels are assembled, the stucco applied to them does not have to be very thick. Most panels are coated with between 2.5 and 3 cm of stucco on each side. The insulation between the sheets of wire mesh may be from 5 cm thick to 30 cm thick. As a result, the inside of the house tends to approach the ground temperature under the house.

There are people who put a conventional roof on a SCIP home, but most such homes are no longer disaster-resistant.

Dave Stevenson is developing a 60-square-meter SCIP kit home that:

1. Can be delivered to the building site on pallets,

2. Is disaster-resistant, as discussed above, and

3. Can be assembled and finished with a crew of nine people in one week.

Jim Farrell has developed the Met-Rock System. It consists of a trailer-mounted jig and press, so the panels can be assembled right on the job site from wire mesh, Warren trusses, and expanded polystyrene. Since Jim is an expert on spraying concrete, he and his staff have designed their system to be easily coated with concrete. The wire mesh has screed points built in so it is easy to apply the right amount of concrete (stucco) to the wall.

A problem with using SCIP in Haiti is that many portions of Haiti have high chlorides in the water and the soil. The high chlorides would lead to oxidation of the galvanized wire mesh used to provide structural integrity. In areas where chlorides are not a problem, SCIP would be a good choice.

A friend of mine, Mark David Heath, is working on technology to build SCIP panels without any steel. My father used to say, "The Impossible Just Takes a Little Longer." So, one of these days Mark will be talking to me about writing a book about his new technology.

Mark David Heath and I are discussing developing technology so that rather than using expanded polystyrene in the SCIP, recycled plastic water bottles will be used.

Structural Concrete Insulated Panels (SCIP) Formula

Structural insulated concrete panels are pieces of Expanded Polystyrene (EPS) plastic between two panels of reinforcing mesh. Keeping those panels apart are trusses. While a 20-cm SCIP may react like a 20-

cm reinforced concrete wall, it has only 2.54 to 4 cm of "stucco" on each side of the EPS filler.

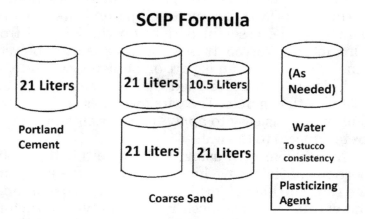

SCIP Formula

Mix cement, sand, plasticizing agent, and water to a mortar consistency.
Sand - Max. particle size 0.8 cm.
Round sand is better than crushed sand.
Discard batches not used within 45 minutes.
Re-temper not more than one time.
Moist cure for 3 days.

Figure 28. SCIP formula

Author's Note: With several others, I am working on a design in which the EPS can be replaced with another material, such as straw, banana leaves, or plastic bottles.

Building Disaster-Resistant Homes

> *Author's Note: With the high chlorides in some parts of Haiti, the SCIP system will not work, since the galvanized steel would be corroded. With several others, I am in the early stages of talking about developing a SCIP which can be built from native materials and will not contain steel.*

With Structural Concrete Insulated Panels (SCIP), vertical rebar need to be placed every 60 cm and alternating to each side of the panels. Preferably, the rebar should be just inside of the reinforcing mesh on each side of the panels. When the panels are finished, the concrete on the outside of them will be 1.0 to 1.5 cm thick. Each rebar needs to be placed into the footing at least 60 cm and must extend up above the top of the stem wall not less than 45 cm. This will allow the SCIP panels to be anchored to the footings. See more information in the section on SCIP.

Insulated Concrete Forms

These are a fun product to work with. Picture large block, often consisting of two sheets of foam insulation board that are held apart by braces. They may be the size of two or four concrete block or even more.

A slab is poured with rebar coming out of the slab every 0.5 meters or less. The first layer of block is usually placed on a bed of mortar, since the slab may not be level. Then additional block are stacked and locked together. Rebar, both horizontal and vertical, are placed in the cavity. Concrete is poured into the cavity, and, if nothing goes wrong, the result is a reinforced concrete wall. If something goes wrong, it is usually the failure of one or more of the block or because the concrete did not fill all the cavities that it should have. The causes are:

- The concrete was poured too fast,
- The concrete was rodded too aggressively,
- The concrete was vibrated too aggressively,
- The concrete was not rodded aggressively enough, or
- The concrete was not vibrated aggressively enough.

If the area where the ICF structure is being built has a chloride problem, the steel rebar can be replaced with either basalt rebar or fiberglass rebar.

ICF Wall

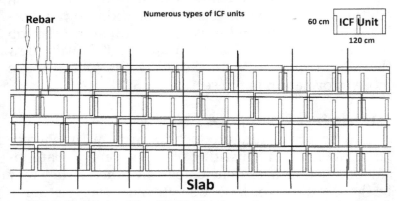

Numerous types of ICF units

60 cm ICF Unit

120 cm

Rebar

Slab

Vertical rebar shown every 120 cm. Every 60 cm is better.
HorizonrL rebar through every ICF unit.
Rebar stubs tied to vertical rebar.
Vertical rebar tied to horizontal rebar.

Fill ICF wall with concrete and rod it to consolidate the concrete.
Plaster outside and inside of wall.

Figure 29. Insulated concrete forms. Picture large block made from 60 cm by 120 cm sheets of Styrofoam. The sheets are held apart about 15 cm. As they are stacked, rebar is placed. Then a concrete slurry is poured in and consolidated.

Insulated concrete forms have some problems:

1. From a thermodynamic standpoint, the insulation is in the wrong place to do the most good.

2. The rebar is in the center of the concrete, so it is not in the ideal location.

3. The outside and inside of the structure needs to be coated to protect it from the elements.

4. If care is not taken, there are voids in the concrete wall.

All that having been said, it is a good building system, and disaster-resistant homes are regularly built using this technology when time is taken to take care of the details.

In summary, insulated concrete forms are sheets of expanded polystyrene (EPS) plastic that are held apart in different ways. Vertical and horizontal rebar are added, and then concrete is poured, using the EPS as the forms. The forms are left in place and act as insulation. The exterior and the interior need to be finished, usually with a stucco.

Insulated Concrete Forms (ICF) Formula

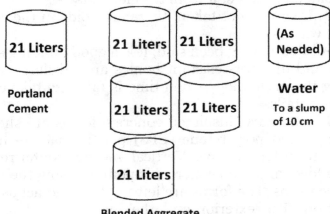

Insulated Concrete Form Formula

21 Liters — Portland Cement

21 Liters | 21 Liters

(As Needed) — **Water** — To a slump of 10 cm

21 Liters | 21 Liters

21 Liters

Blended Aggregate

Mix cement, blended aggregate, and water to a slump of 10 cm.
See Chapter 4 for information concerning blended aggregate.
Maximum aggregate size is 1/3rd the distance between the EPS forms.
If a well-blended aggregate is not available, more cement is needed.
Pour, consolidate, screed, and rough float.
Moist cure the top for 3 days.
Leave the EPS forms in place.

Figure 30.   ICF formula.

Confined Masonry Construction

The most common method of concrete block construction in Haiti and many parts of the third world is known as Confined Masonry Construction. Following the information provided in Chapter 10 on masonry construction does not result in a disaster-resistant

Building Disaster-Resistant Homes

structure, but by incorporating that technology into a con fined masonry building, disaster-resistance can be achieved.

After laying out the building and before pouring the foundation, vertical rebar columns need to be placed. Each column consists of four No. 3 rebar that are joined at specific intervals by No. 2 rebar stirrups.

The rebar stirrups should be placed at each corner and at critical points within the structure. Critical points include at each location where two walls intersect, where a wall ends, and on each side of each doorway and window. Additionally, the rebar columns should not be placed more than 4.5 meters apart.

When laying up the units in the wall, they should be laid with a running bond pattern, and each end should be toothed so the concrete which is poured will have a better bite. After the wall is laid, forms are added around each rebar truss, and concrete is poured and rodded to fill the space.

Beams need to be poured about halfway up the wall (just below the windows) and again at the ceiling/roof line. These beams should be reinforced with not less than four pieces of No. 4 rebar which are tied at specific intervals with No. 2 rebar stirrups.

Quality Concrete from Crap

Toothed Column for Confined Masonry

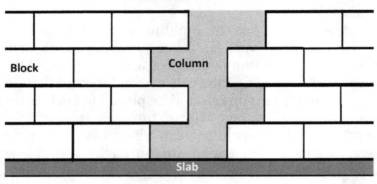

Block Column

Slab

Figure 31. Toothed column for confined masonry. Do not use half block or cut block. The width of the column at each tier of block needs to be a full block wide. A rebar truss needs to be placed in the column.

The beams in conjunction with the roof slab need to be poured monolithically, and the rebar in the beams and the rebar in the roof need to be tied together, then the roof needs to be poured monolithically. (See the section on lightweight concrete roofs for an alternative to pouring the entire roof monolithically.)

The problem with confined masonry in parts of Haiti with high chloride levels is that steel rebar is used for reinforcing. When the rebar oxidizes, the structure fails.

I developed a truss fabricated from basalt rebar and basalt rope to be used in columns and in beams. It is triangular rather than square in cross section, so it is a little more difficult to get good penetration of the concrete into the beams and columns. I have not had the time to develop a jig for forming square truss-

Building Disaster-Resistant Homes

es from basalt rebar and basalt rope, but that will come.

Cisterns

While a cistern will not protect the house from a hurricane, it does protect the occupants from one of the major effects of hurricanes: the shortage of fresh water. A water collection system (with filters) built into the home can provide a family with fresh water for many days, if not weeks, if used frugally. If built on an outside wall but inside the home itself, the water is fairly well protected. It would still require disinfecting, but that is a relatively easy process. (Two drops of chlorine per liter of room-temperature water requires a twenty-minute wait and will yield palatable [tastes ok] and potable [safe] drinking water.)

Chapter 15
Measuring Strength

Why ASTM?
American Society of Testing and Materials (ASTM) standards were developed so buyers, sellers, engineers, and others could know specifically what was being bought, sold, used, specified, etc. In recent years the standards have gone worldwide so international trade can be enhanced.

There are ASTM procedures to follow, and special equipment is needed to measure the strength of concrete. The ASTM system has been developed so that no matter where a product is tested in the world, all the results will be very close. As a result, the process is expensive. If certified results are needed, this is the only way to go.

This chapter explains how, for a relatively reasonable amount of money, one can get a **reasonable estimate** of the strength of a concrete, a stucco, or a mortar. What I would call a guesstimate.

When working for Pozzolana, Inc., in the 1950s, we needed to test the compressive strength of cement mortar with the pozzolan we were producing. One of the quality control tests was to mix the mortar paste, place it in wooden molds we had made and soaked in waste motor oil, cure the samples for 3 and 7 days, and then break the cubes. The press we used was a hydraulic jack that was mounted in a frame. A pressure gauge was attached to measure the oil pressure. A needle was advanced as the pressure increased and remained in place after the sample broke until we re-set the needle to zero. It served our needs.

When the lab upgraded to metal molds which were tinned, a series of tests were run comparing the wood molds to the metal molds and found that higher compressive strengths were obtained from the wood-en molds. Probably the cause was that the wooden molds absorbed some of the moisture in the paste, which gave the paste a better water/cement ratio; and then as the concrete cured, that water was released back to the samples keeping them at an optimum moisture content for curing.

ASTM requirements for compressive test ma-chines and molds were enhanced every year or two. We ended up being required to use brass molds which cost about a week's wages and had to be recertified every few years.

At one point a hand-pumped compressive test machine was acceptable, but now the rate of increase in pressure needs to fall between very tight limits. The reason – Increasing the pressure rapidly can give higher readings. Increasing the pressure very slowly can give lower readings. The standards are needed, they are good, and they are expensive.

A Nail, a Credit Card, & a Magnifying Glass

Most people who do testing of mortars, stuccos, and other concretes are asked to go out into the field and figure out why a job went bad. For the last 40 years when inspecting stucco and mortar jobs, I usual-ly carry a 16d nail, a 10X magnifying glass, and a credit card.

If well-cured stucco or mortar can be scratched with a nail, the material is very weak. If it can be marked with the nail, but not scratched with it, it has an appropriate strength. If the mortar can be marked but not scratched, it probably has a compressive strength of about 10 MPa.

If rubbing a finger over a crack indicates displacement, the problem is more complicated than a drying shrinkage or plastic shrinkage crack.

Most credit cards are about 0.075 cm in thickness. If there is a crack that the credit card can just barely fit in, one can return to see if the crack has expanded if the area is marked. A credit card will not go into most drying shrinkage and plastic shrinkage cracks.

The magnifying glass allows one to look for signs of efflorescence, staining, etc., and to inspect the bond between the masonry unit and the mortar. The most important use of the magnifying glass is to intimidate the contractor. Many contractors will answer questions based on how they know the job should have been done, not on what was actually done. If one exams a wall for 5 minutes with a magnifying glass and says, "Hmm," and comments, "Strange," several times, and then records something in a notebook, the contractor will usually be nervous, and then provide honest answers to questions.

Molds

Molds come in several different sizes. They are available with the English system of inches and with the metric system of millimeters. For simplification, we will address only the metric system.

Common cube molds are 50 mm and 150 mm. Cements, mortar, and stucco would be tested in the smaller molds, and concrete with aggregate up to 50 mm would be tested in the larger ones. These are heavy metal precision molds.

Plastic cylinders are available for making concrete cylinders to test. Standard sizes include:

5-cm x 10-cm

10-cm x 20-cm

15-cm x 30-cm

The maximum aggregate size is one-third the inside diameter of the plastic cylinder.

Compressive test strengths are meaningless unless dimensions and ratios are specified. The larger the mass of the concrete, the higher the strength. The taller the specimen related to the cross-sectional dimensions, the lower the compressive strength.

Always stay with the accepted sizes of the cubes or cylinders.

Author's Note: Years ago, one of my employees tested a concrete cylinder without taking it out of the plastic cylinder mold; that increased the apparent strength considerably.

Filling the Molds

With concrete in the field, a slump test is usually used to determine the consistency. It is also an indication of the amount of water that is in the mix, but if the mix is allowed to set for a while, the mix will get stiffer and the slump will decrease.

The slump test device is a truncated cone 30 cm in height with both ends open. Concrete is added and rodded, and then struck off level at the top. The slump cone is removed, and the amount that the concrete slumps without the support of the cone is the slump. It is usually measured in inches. Concrete with a slump of 1 inch would be very stiff. With a slump of 8 inches, it would be very soupy.

For mortars and stuccos, a flow test is used, since most mortars and stuccos are designed to hold their shape. It is essentially a variation of the slump test. A 5.0 cm high truncated cone is filled, rodded, and struck off. The cone is removed, and the table the sample is sitting on (4.08 kg) is lifted slightly (1.27 cm) and dropped 25 times. The flow is the percent that the base diameter of the specimen expands. The flow table is expensive, and it is not likely that one could fabricate a low-cost replica which would be close to accurate.

The molds are then filled in lifts and rodded in a specific fashion. For 50 mm cubes, the mortar needs to be tamped 32 times to ensure that it is well-consolidated.

If these preliminary tests were not required, people could reduce the amount of water for the tests and have results that show that the concrete had much greater strength than it actually had.

Compression Test Machine

A compression test machine which will give reasonable results can be built for a fraction of what an accurate and certified machine would cost to purchase. By building and using one of these, many decisions can be made before paying for certified tests.

Protection

First off, when a cube or a cylinder breaks, often chunks of concrete are thrown with great force. The higher the compressive strength, the greater the force that they can be thrown. Build a protective barricade before an eye is lost.

Force Needed

Decide on what will be tested and the anticipated strength of the samples.

Example: to break 50-mm cubes that have a maximum compressive strength of 5,000 psi, 20,000 pounds of pressure would be needed. That would require 10 tons of force.

Example: to break 100-mm cylinders that have a maximum compressive strength of 5,000 psi, 62,800 pounds of pressure would be needed. That would require 32 tons of force.

If 20 tons of force are needed, a bottle jack may be more economical than a port-a-power. When higher forces are needed, then a port-a-power may be the only option.

The Jack Stand

The jack stand needs to have a stable base and needs to be tall enough so the power source and the sample can be inserted. Ideally, the space where the sample is placed to be tested should be adjustable.

> *Author's Note: Extra space comes in very handy. I have a test machine that I had fabricated for testing 50 mm cubes and 50 mm x 100 cylinders. Then I decided to test the pull-out strength for screws and bond strength. This required my machinist to build some jigs for me which were taller than what I ever dreamed of needing. My machinist had anticipated that I would want to do something weird and had added an extra 15 cm of space that I had not requested.*

Measuring Strength

If the upper and lower surfaces (called platens) which compress the test specimen are not perfectly flat against the upper and lower surfaces of the sample, the tested compressive strength will be lowered. This is taken care of by placing a steel ball, from a ball bearing, above the upper platen and another below the lower platen. Ideally the steel balls should be about 1-cm in diameter.

To keep them in place, a shallow hole can be drilled in the center of each platen.

The platens should be thick enough so they do not bend or deform when pressure is applied.

The oil pressure of the jack or the port-a-power needs to be measured. A pressure gauge can be inserted into a special port on some of these units. This is not the oil fill port. The gauge needs to be sized for the anticipated oil pressure. This is done by taking the maximum pressure and dividing by the cross section of the jack cylinder.

Example: With a 10-ton jack (20,000 lbs.) and a cylinder of 1 inch, the cross section would be 0.785 sq. in., and the oil pressure at that load would be 25,477 psi. If the cylinder were 2 inches in diameter, the oil pressure at that load would be 6,370 psi. If there is a choice, go with the largest diameter cylinder available.

A digital gauge is easier to use and usually more accurate. If a regular pressure gauge is used, look for the largest diameter one that is available and has a needle that will show the maximum pressure recorded.

Chapter 16

How Long Will This Concrete Job Last?

Go back to the Prologue and read the questions and answer them again. If you want to be a quality builder and to build homes which will last past the next hurricane or earthquake, you need to apply what you have learned in this book.

Each individual building his own home should ask, "If I don't have the time or money to do it right in the first place, will I ever have the time and money to fix the problems I created?"

Each professional builder should ask, "If there are not enough resources to build a quality home for a client, should I even consider taking the job since I will be causing that client problems over the years and may kill him and his family with the next disaster?"

In the final analysis, build quality homes because that is what your clients need.

Addendum A

Conversion Chart—Page 1

Unit	Symbol	Times	Equals	Symbol
Volume				
Pint	pt	0.473	Liter	l
Quart	qt	0.946	Liter	l
Gallon	gal	3.785	Liters	l
Cu. inch	in³	16.4	Milliliters	ml
Cu. foot	ft³	28.32	Liters	l
Cu. foot	ft³	0.02832	Cubic meter	m³
Cu. yard	yd³	0.7646	Cubic meter	m³
Length				
Inch	in	25.4	Millimeters	mm
Inch	in	2.54	Centimeters	cm
Foot	ft	30.5	Centimeters	cm
Yard	yd	0.9144	Meter	m
Mile	mi	1.61	Kilometers	km
Area				
Sq. inch	in²	645	Sq. millimeters	mm²
Sq. foot	ft²	0.0929	Sq. meter	m²
Sq. yard	yd²	0.836	Sq. meter	m²
Acre	ac	0.4047	Hectare	ha

Conversion Chart—Page 2

Unit	Symbol	Times	Equals	Symbol
Density				
Ounce	oz	28.35	Grams	g
Pound	lb	454	Grams	g
Pound	lb	0.454	Kilogram	kg
Ton (short)	tn	0.907	Tonne	Mg or t
Pressure				
Pounds / sq inch	psi	6.895	kilopascals	kPa
Pounds / sq inch	psi	0.006895	megapascal	MPa
Temperature				
Degrees F	F°	(F°-32) x 0.556	Degrees C	C°

Addendums

Addendum B

The following topics need to be addressed in separate books, booklets, or papers. Who will volunteer to start organizing and writing?

Lining Sewers with Concrete

Sewage contains organic wastes. Decomposition of organic wastes develops acidic conditions. Acids attack concrete. Most acid-resistant concrete formulae do not use Portland cement as a base, because when Portland cement hydrates, it gives up calcium ions which are an acid's favorite food.

> *Author's Note: A noted scientific laboratory tested an acid-proof concrete and determined that it would last for 700 years if it were made as concrete pipe and used to carry human sewage. My testing (I used a different protocol) showed that it would probably last for about 7 years. When the product was made and used, deterioration was noted in 2 years. It always helps to have a test that has been proven to be accurate.*

Following are steps which can make concrete made from Portland cement more acid-resistant.
• Do not use aggregate which is soluble in acid. Soft limestone is the worst aggregate which can be used.
• Any limestone aggregate is bad.

- Ensure there is enough cement paste to fill all the voids between the aggregate particles.
- Use a pozzolan, and chemically balance the formula so that all calcium ions produced are tied up with the pozzolanic reaction.
- Keep the entrained air as low as possible.
- Use as little water in the mix as possible, but enough to get a good bond to the aggregate.
- Consolidate the concrete as much as possible.
- Use a long-lasting water repellent.
- Do not re-temper the concrete if it starts to stiffen before being used.
- Mechanically mix the concrete.
- Do not attempt to make acid-resistant concrete unless there is a reliable concrete laboratory available to help develop the formulae and available to provide quality control.

Ferro-cement—Concrete Without Forms

Traditional Ferro-cement was several layers of wire mesh bound into an armature and then stuccoed. A plasterer would be on each side, and they would try to force the stucco into the mesh between them. Some ships used in WW II were built with this technique. Then water-reducing agents and plasticizers were developed so a stucco mix that would develop great strength could be sprayed. Now sculptures, water tanks, thin columns and shapes, and much more, are made from Ferro-cement.

Meet a 13-foot dinosaur which stands in front of the Witte Museum in San Antonio, Texas. He was built in a week as a project for a decorative concrete show.

With a curve in a Ferro-cement panel, it becomes stronger. With a double curve, it becomes even stronger. In the US, engineering students at different colleges participate in a concrete canoe competition. These canoes are not necessarily practical boats, but they challenge the students to think of ways to build strength while using as little material as possible.

Author's Note: Several of my friends and I have consulted with several of these teams. One of the regional winners found that using Spider-Lath (a fiberglass cloth) gave much greater strength than metal lath.

Concrete Block Manufacture

The book should address:
- Mix design
- Block sizes
- Vibration
- Cure
- Testing
- Storage

To test the quality of a block, drop it from chest height onto a grassy area. If it breaks, the block are not strong enough for building. Caution: hold the block away from the body so toes do not get smashed.

Accepted Practices and Standards

There are many practices and standards concerning the production and use of concrete around the world. They have not been included in this book because the intended user of this book will not have them available and may not have all the materials available to use them. Some of the more important organizations/documents would be:

International Code Council, Inc. (ICC) The ICC is a leading authority and resource for developing building codes. Building codes were developed to ensure that all buildings were built in a safe manner. Besides codes which address structures, there are codes which address plumbing, fire, electricity, and energy. Most codes are updated every three years. Two of the more important ICC codes are:

International Building Code. (IBC)

International Residential Code. (IRC)

National Concrete Masonry Association. (NCMA) The NCMA is the national trade association representing the producers and suppliers of concrete masonry products, including concrete block, manufactured stone veneer, segmental retaining walls, and articulating concrete block.

American Concrete Institute. (ACI) The ACI is a leading authority and resource worldwide for the development and distribution of consensus-based standards, technical resources, educational programs, and proven expertise for individuals and organizations involved in concrete design, construction, and materials, who share a commitment to pursuing the best use of concrete.

American Society of Testing & Materials International. (ASTM) ASTM is an international standards organization that develops and publishes voluntary consensus technical standards for a wide range of materials, products, systems, and services.

Addendum C
Research Needs

Can nopal be used as a substitute for hydrated lime in mortar and stucco? When I was a kid, my father sometimes burned the spines off prickly pear cactus, diced it, crushed it, mixed it with water, and let it soak for two days. He added it to a Portland cement/sand mix for making mortar and stucco. I need to repeat his process so I can determine the appropriate ratios. He always said that we should not make up more than was needed, because after a few days it started stinking. It gave body to stucco and re-inforced it.

Can guar be used as a substitute for hydrated lime in mortar and stucco? Kel-Crete is derived from an extract from guar bean. Can guar be grown in Haiti and other third world coun-tries and then crushed and minimally processed so it can be used as an additive to mortar and stucco to replace hydrated lime? If it works, micro businesses could be developed to grow and process the guar.

Can plastic bottles be cut and used to replace basalt rope? Using a rotary cutter, strips of plastic which are 0.32 cm wide and about 5 meters long, could be formed into rope and used as a replacement for basalt rope. If such rope can be developed, then it would be necessary to test to determine whether such rope would be

*strong enough to reinforce domes. If it is fea-
sible, micro businesses could be developed to
make plastic reinforcing rope.*

*Can plastic bottles be used to replace EPS in SCIP?
EPS is used to support the wire mesh and the
Warren trusses in SCIP panels. Currently, if
the EPS is removed, the panels will collapse. Is
there a way to stabilize the panels and then fill
them with plastic bottles to provide the insulation for the panel? It probably would not be as
effective as EPS, but it would be using a waste
material in Haiti and other countries. It would
probably be necessary to insert burlap bags or
something similar to hold the bottles to reduce
the amount of concrete which would have to
be sprayed on the panels.*

*Can a non-metallic frame be built for SCIP? In many
places, SCIP is an excellent building material,
but in some areas, such as the lowlands of
Haiti, the chloride content is high, and galvanized steel tends to corrode and lose its
strength. Is it possible to build a non-metallic
SCIP frame? Bamboo is one alternative. Basalt
mesh to replace the metal mesh is another alternative, as the price of basalt drops.*

*Can bamboo be used effectively for reinforcing con-
crete? The answer is "yes," but more work
needs to be done, especially on the amount of
bamboo needed and the resistance to deterioration where moisture is present.*

*Can a method be developed for tying the EcoShell
Airform to the concrete slab which does not
require drilling into the slab and using Tapcon*

Quality Concrete from Crap

Anchors. A problem with the present method is that even if the slab is adequately marked, the team tends to not follow the marking and the dome ends up not being round. Removing the Airform is difficult when removing the Tapcon anchors, and in some cases a grinder needs to be used to cut the head of the anchor. This can lead to damaging the Airform.

Can SpiderLath be used for making ferro-cement roof channels? How many layers of Spider-Lath are needed to give the equivalent strength of steel-lath-reinforced ferro-cement roof channels?

Addendums

Addendum D
Glossary

Term   Description

Abram's Law: This law states that the strength of a concrete mix is inversely proportional to the ratio of absolute volume of water to the absolute volume of cement. As the water content increases, the strength of concrete decreases.

Accelerators: A cement accelerator is an admixture used to accelerate the set of cementitious products. It may or may not accelerate the cure. (Set time is based on the speed that hardness is achieved. Cure is based on the chemical reactions which take place within the paste.)

Aggregate: Aggregates should be inert granular materials such as sand, gravel, or crushed stone and are usually sized through screens to specific sizes. Optimally, the amount of each component should be such that the resulting aggregate will have a minimal pore space and thus will be as dense as possible.

Air/Cement Ratio: See Abram's Law.

Basalt Rebar: Rebar formed from tiny threads of basalt filament which are held together with a binder such as epoxy.

Basalt Rope: Rope which is formed from tiny threads of basalt filament which are twisted into rope.

The rope is used as reinforcement in concrete and is not suitable for the normal purposes for which rope is used.

Bagasse Ash: Bagasse is the residue from sugar refining. It is burned as biofuel, and the resulting ash is a pozzolan. It can be used to replace up to 20% of the Portland cement in a concrete mix.

Cement: A substance which when mixed with water and aggregates hardens to a rock-like consistency.

Cement Paste: A mixture of cement and water.

Chloride(s): An anion which forms very soluble compounds. It functions as a strong accelerant when mixed with cement pastes. Since it causes deterioration of steel in concrete, its use in concrete has been severely limited.

Columns: Vertical concrete structures normally used for support or bracing. They are often poured between wall panels of concrete block which have been laid. See Wall Beam, Roof Beam, and Confined Masonry Construction.

Compressive Strength: The load an object will withstand. Usually measured as weight per given volume. MPa or psi.

Concrete: A mixture of cement, water, and various sizes of aggregate which will harden with time. It may contain one or more admixtures to modify the properties of the concrete.

Addendums

Concrete Slump Test: The concrete slump test or slump cone measures the consistency of fresh concrete before it sets. It is performed to check the workability of freshly made concrete, and therefore the ease with which concrete flows. It can also be used as an indicator of an improperly mixed batch.

Confined Masonry Construction: A building system where panels of masonry are surrounded by beams and columns to give them strength.

Entrained Air: Tiny bubbles of air mixed into the concrete.

Expanded Polystyrene: Polystyrene with entrained air to give it a density of 1 to 7 pounds per cubic foot. It is an excellent insulator.

False Set: A false set is when non-cementitious reactions, often involving gypsum, cause the mix to appear to set, but no actual chemical hydration of the cement particles has taken place.

Ferret's Law: This law states that the strength of a concrete mix is inversely proportional to the ratio of absolute volume of water plus the absolute volume of air to the absolute volume of cement. As the water content and the air content increase, the strength of concrete decreases.

Ferro-cement: A structural system where a sanded cement paste with a low water content is forced into an armature made of multiple layers of wire mesh.

Fines: Sand-sized material – from 0.25 mm to 3.2 mm.

Fly Ash: Fly ash is a by-product of burning coal. Normally it is classified as Type F (usually high in pozzolanic activity) and Type C (usually contains up to 50% low-grade cement chemicals which provide rapid strength growth but contributes less than Class F to the long-term strength of the product).

Form(s): Barriers, often wooden or steel, which hold concrete in place as it is poured and as it cures.

Hydrated Lime: Calcium hydrate. It is formed by calcining limestone and then adding measured amounts of water to hydrate the resulting calcium oxide. Normally, it is classified as Type S (dolomitic and highly hydrated), Type N (high calcium and not as highly hydrated), or industrial grade (not as highly hydrated, and the particle size is not as controlled).

ICF: Insulated Concrete Forms – A building system which uses forms that are made from EPS or similar material and stay in place after the concrete cures. Walls are more insulative than many standard walls, but not as insulative as structures where the insulation is placed to the outside of the structure.

Igneous Rock: Igneous rock were at one time molten. Besides the difference in the chemical makeup of the magma (lava), the speed of cooling has a great impact on the characteristics of the rock. Basalt is formed when a siliceous magma cools rapidly and becomes amorphous (without crystals); granite is formed if the magma cools slowly and numerous crystals are formed.

Jitterbugging: Vibrating the surface of a concrete slab to consolidate the concrete.

Kel-Crete: A common but excellent mortar fat made of a derivative of guar beans.

Lath: A substance used as a base for stucco or plaster. Originally it was thin wood slats, then metal lath was developed. In recent decades fiberglass lath has been developed. The lath may be flat, ribbed, or distorted so it will be about 0.6 cm from the substrate.

Lime: Any of several forms of hydrated lime (calcium hydroxide). Some people call ground limestone "lime."

Metakaolin: Clay that has been heat-treated to remove some of the chemically combined water and increase its pozzolanic activity.

Metamorphic Rock: Metamorphic rock are usually sedimentary rock that have been changed by heat and/or by pressure. The marble in Haiti started as limestone and then gradually changed to marble when exposed to both heat and pressure.

Met-Rock System: A form of SCIP which contains ridges in the face mesh to act as screed points.

Mold(s):   Forms constructed for use in forming mortar, stucco, and concrete samples for testing in the laboratory.

Mortar: A mixture of cement, sand, water, and other components to be used between masonry units. Mortar has two primary functions: to hold the masonry units apart, and to provide bond strength so the structure is stable.

Mortar Fat: Mortar fat is a name that is applied to many different products which are used in stucco and mortar as a replacement for hydrated lime. They are also called plasticizing agents.

Non-Reactive Fines: Clay, rock dust, soil, or anything else which is fine in nature and does not participate in the cement-curing reaction. As a result, they can lower the compressive and tensile strength of products formed from the cement.

Nordmeyer's Law: This law states that the strength of a concrete mix is inversely proportional to the ratio of absolute volumes of water, plus air, plus non-reactive fines, to the absolute volume of cement. As the water content, the air content, and the non-reactive fines content increase, the strength of concrete decreases.

Papercrete: A product formed by slurrying paper and cardboard. Often it contains entrained air to make it more insulative. It can be made into block for building, plastered into place, or poured to fill cavities.

Particle Size Distribution: A list of the percentage of each of the different-sized particles in an aggregate.

Plastic: Can be formed, such as moist clay or fresh concrete.

Plastic Shrinkage: The shrinkage of a mortar, stucco, or concrete as it cures and as any excess mix water surfaces and evaporates.

Platens: Normally, thick steel plates which are placed above and below a sample to be tested in a compressive test machine.

Plinth: A plinth is traditionally the base of a column. See Stem Wall.

Portland Cement: Any of several grades of cement which meet ASTM Standard C150. This is the most common cement on the market.

Pozzolan: A non-crystalline and finely divided amorphous (non-crystalline) silicate or aluminate which in the presence of calcium ions can form hydrated calcium silicates and aluminates (hydrated cement chemicals).

Quick Lime: Calcium oxide – The result of calcining limestone and one step in the production of hydrated lime.

Rebar: Rods of reinforcement for use in concrete. Steel is the most common material used. Sizing numbers refer to the diameter of the rod in eighths of an inch. A No. 3 rebar is 3/8 inch in diameter. The rebar may be coated with paint, epoxy, or other substances to protect it from corrosion. Besides steel, the rebar may be made from fiberglass, bamboo, or basalt fibers.

Rebar Stirrups: Rebar stirrups are used in beams and columns to hold the rebar in position and to some extent share the stresses between the different rebar pieces.

Reinforcement: Anything that adds tensile strength to cementitious products, ranging from rebar, to wire mesh, to lath, to fibers.

Retarders: An admixture which slows the set of cement chemicals. Use of cold water is one of the most common retarders.

Rodding: Using a rod to consolidate the concrete in a column or in a mass pour. See Jitterbugging.

Roof Beam: See Column, and Wall Beam.

Rotary Kiln: A rotary kiln is a long cylinder rotating slowly about its axis. Usually the burner is on the lower end and there is a slight downward incline from the feed end to the burner end. The concept is to heat all the feed an equal amount and keep it stirred so it does not consolidate if the calcining temperature is close to the melting point of the feed.

Rule of 4: With spherical objects, the densest mix can be achieved if, starting from the finest particles, each next-sized particle group is 4 times the diameter of the previous group. Example: 1/32″, 1/8″, ½″, 2″, 8″

SCIP: Structural Concrete Insulated Panels – A building system where a solid insulation is enclosed between two sheets of wire mesh, and the space between the two sheets of wire mesh is reinforced with angle wires or Warren trusses. After the panels are placed, both sides are plastered with 2.5 cm to 5.0 cm of cement stucco or shotcrete. The resulting wall reacts to stresses in a way that is similar to a reinforced solid concrete wall of similar dimensions.

Sedimentary Rock: Sedimentary rock are formed from material settling in water, or from the air, or from being precipitated chemically from solution. In time, they solidify enough to form rocks. Limestone is the most common sedimentary rock found in Haiti.

Shotcrete: A plaster or concrete with small aggregate that is placed by propelling it from a "gun" with air. Usually, the plaster is delivered to the "gun" with a pump. The plaster ends up being better consolidated than if applied with a trowel. Gunite is the brand name of one of the first shotcrete systems and often used interchangeably with the term shotcrete.

Silica Fume: Finely divided silica that is formed during silicon chip production. It is an excellent pozzolan, but must be used with water reducers and other admixtures.

Slab: A piece of flat horizontal concrete. It may be on the ground, elevated to form an upper floor, or elevated to form a roof.

Slump Cone: A slump cone is a truncated cone that is 12 inches high, with a base diameter of 20 cm and a top diameter of 10 cm. The cone is filled with fresh concrete in three layers of equal volume. Each layer is rodded 25 times with a rod that is ¾ inch in diameter and has a rounded end. After the final rodding, the surface is screeded off and the cone lifted off. The slump is the distance the concrete has slumped when not supported by the cone.

Stem Wall: The stem wall is traditionally a wall that connects a footing to the above-ground portion of a building. See Plinth.

Stucco: A mix of cement, sand, water, and admixtures, which is designed to be troweled or sprayed on a wall or a ceiling to provide a smooth or a textured surface.

Tensile Strength: The greatest stress a substance can bear without being torn apart. Since cement products are strong in compression and weak in tension, reinforcing is added to increase the tensile strength.

Truss: A fabricated unit which serves as support or as reinforcing. For roofs, trusses replace rafters. For concrete columns and beams, trusses are units which tie several pieces of rebar into a specific shape and hold them in that shape when the concrete is poured and rodded around them.

Void: An area in concrete which does not contain aggregate, reinforcement, or cement paste.

Addendums

Wall Beam: See Column, and Roof Beam.

Warren Truss: A Warren truss is a support structure used in different constructions, for supporting a load. It consists of an upper chord and a lower chord. In between are members placed at an angle to form a series of triangles. Since the internal bracing resembles a W, they are often called W trusses.

Water/Cement Ratio: A ratio between the mass of the water in a mix and the mass of the cement in the mix. Normally the higher the ratio, the lower the strength of the product. See Abram's Law.

Water-Reducing Agents: Admixtures which reduce the amount of water needed to allow a mortar, stucco, or concrete to become workable. Since many water-reducing agents entrain air, they do not necessarily increase the product's strength. See Ferret's Law.

Wire Mesh: Sheets of wire that are used to reinforce.

About the Author

Herb Nordmeyer was homeschooled in chemistry and cement chemistry. Even though he went on to earn degrees in biology, chemistry, and aquatic ecology, he considers his real education that home-schooling. After following a career track based on his formal education for a few years, he got back into the construction materials field where he specialized in product development and forensic analysis of failures (usually, but not always, other people's failures). Since his retirement in January, 2010, he and his wife own and operate Nordmeyer, LLC, a consulting, writing, and publishing company. He is a partner in HerbCrete, LLC, a company which develops specialty stuccos.

He has been very active in ASTM, serving on Committees: C 1 (Cements), C 11 (Gypsum and Related Products – including stucco), C 12 (Mortar for Unit Masonry), C 15 (Masonry Units), and C 27 (Precast Products).

In September, 2013, Herb made what he thought was his first and last trip to Haiti. Pastor Benoit showed him the need for earthquake-resistant and hurricane-resistant housing. He challenged Herb to find or develop technology to build earthquake- and hurricane-resistant housing for Haiti, especially the slum called Jubilee, which could be built for a materials cost of $1,000 US. In November, 2013, Herb was back in Haiti where he agreed to go at least four times per year, as long as his health and finances held up.

Besides teaching in the slums, in January, 2017, he taught a 4-semester-hour course in disaster-resistant construction at the American University of the Caribbean in Les Cayes, and plans on being back in January, 2018, to teach another course.

Herb is a prolific author, but a poor writer and speller. His wife, Judy, must drip red ink over everything he writes. Besides numerous peer-reviewed papers in ASTM and other scientific journals, he is the author of **Stucco Handbook for Builders** (out of print), and **The Stucco Book–The Basics** (2012).

Homes for Jubilee and the Haitian Creole edition, **Kay pou Jubilee,** interfered with the publication of **The Stucco Book** series of books, and this book has continued that interference.

When not involved in building, or in building materials, or in writing about them, Herb adopts granddaughters and with several of them has published several books; and he enjoys taking people kayaking.

About the Author